Expectation: The Mathematical Framework of Reality

Subtitle: How the Science of Expectation Shapes Our World Through Universal Laws

Detailed Outline

Introduction

Subtitle: Bridging Thought, Intention, and Expectation

- **Purpose of the Book:**
  - Why understanding expectation as a universal force matters.
  - Its role as the bridge between thought/intention and manifested reality.
- **Connection to the Unified Field Theory (UFT):**
  - How expectation completes the triad of thought, intention, and manifestation.
  - Builds upon the principles established in *Harmonic Thought*.
- **What Readers Will Learn:**
  - A groundbreaking framework for modeling expectation mathematically.
  - Techniques to align their expectations with universal laws to shape reality.

Part 1: The Nature of Expectation

Chapter 1: The Science of Expectation

Subtitle: The Role of Anticipation in Human Experience

- **1.1 The Cognitive Foundations of Expectation:**
  - How the brain predicts outcomes based on past experiences.
  - The role of neural pathways in shaping anticipatory behavior.
- **1.2 Evolutionary Advantage of Expectation:**
  - How expectation evolved as a survival mechanism.
  - Connection to reward systems and decision-making.
- **1.3 Expectation in Psychology and Behavior:**

- Placebo and nocebo effects as evidence of expectation influencing reality.
- Experiments showcasing the power of belief in shaping outcomes.

Chapter 2: Expectation and the Ether

Subtitle: Resonance and the Universal Substrate

- **2.1 The Ether and its Role in Reality:**
  - Revisiting the Ether as a dynamic substrate for universal forces.
  - How thought waves and expectations propagate through the Ether.
- **2.2 The Quantum Connection:**
  - How expectation aligns with quantum phenomena like wave-particle duality.
  - Double-slit experiments and observer effects as evidence.
- **2.3 Aligning Expectation with Universal Patterns:**
  - How the Golden Ratio (ϕ) serves as the harmonic constant.
  - Resonance as the key to aligning expectation with physical reality.

Part 2: The Mathematical Framework of Expectation

Chapter 3: Modeling Expectation Mathematically

Subtitle: From Intention to Anticipation

- **3.1 The Evolution from Intention to Expectation:**
  - How intention focuses energy, while expectation directs outcomes.
  - The feedback loop between anticipated outcomes and intention refinement.
- **3.2 The Expectation Equation:**
  - Deriving expectation mathematically:

$$E_x = \int_{t=0}^{T} \left(I(t) \cdot P(t) \cdot \phi \right) dt$$

  - Defining variables:
    - $I(t)$: Intention's strength and coherence.

- $P(t)$: Predictive resonance, or alignment between thought and universal patterns.
- \phi: The amplifying harmonic constant.
- **3.3 Feedback Loops and Resonance Cycles:**
  - Introducing iterative models for expectation calibration.
  - How reality strengthens or weakens future expectations.

Chapter 4: Testing the Framework

Subtitle: Validation Through Experimentation and Observation

- **4.1 Empirical Evidence for Expectation's Power:**
  - Psychological experiments (e.g., random number generators, placebo effects).
  - Historical examples of collective expectation shaping societal outcomes.
- **4.2 Proposed Experiments to Test Expectation:**
  - Testing predictive alignment in physical systems.
  - Exploring expectation's impact on quantum states and biological responses.
- **4.3 Practical Validation Techniques:**
  - Methods to measure predictive resonance and coherence.
  - How to observe feedback loops in real-world contexts.

Part 3: Applications of Harmonic Expectation

Chapter 5: Personal Transformation

Subtitle: Mastering Your Expectations to Master Your Reality

- **5.1 Reframing Expectations:**
  - Techniques to shift negative patterns into positive anticipatory states.
- **5.2 Coherence and Predictive Resonance:**
  - How to align expectations with universal harmony for consistent outcomes.
- **5.3 Harnessing the Feedback Loop:**
  - Practical exercises for refining expectations based on outcomes.

- **5.4 Healing and Growth Through Expectation:**
  - Case studies of physical and emotional healing via positive anticipation.

Chapter 6: Societal Impact of Collective Expectation

Subtitle: Shaping the Future Through Unified Vision

- **6.1 The Power of Collective Expectation:**
  - Historical and modern examples of societies aligning toward a common goal.
- **6.2 Social Movements and Resonance:**
  - How shared expectations ripple through societal systems.
- **6.3 Governance and Policy Informed by Expectation:**
  - Implementing expectation-based decision-making models.
- **6.4 Harnessing Global Resonance:**
  - Techniques for uniting humanity under shared positive expectations.

Part 4: The Future of Expectation and the UFT

Chapter 7: Expectation in the Unified Field Theory

Subtitle: A New Force in the Universal Framework

- **7.1 Completing the Triad:**
  - Thought, intention, and expectation as interconnected universal forces.
- **7.2 Expectation as Predictive Resonance:**
  - How expectation interacts with matter, energy, and the Ether.
- **7.3 Implications for Science and Philosophy:**
  - Redefining human agency in shaping the cosmos.

Chapter 8: A New Paradigm for Reality

Subtitle: Harnessing Expectation for Individual and Universal Harmony

- **8.1 Humanity's Role in Universal Evolution:**
  - How expectation advances collective consciousness.

- **8.2 Beyond the Physical: The Metaphysical Implications:**
  - How expectation bridges the gap between science and spirituality.
- **8.3 The Legacy of Expectation:**
  - Shaping a future where humanity harmonizes with universal laws.

Introduction

Subtitle: *Bridging Thought, Intention, and Expectation*

Purpose of the Book

For centuries, humanity has sought to understand the forces that shape our reality. From ancient metaphysical traditions to cutting-edge quantum physics, we have probed the boundaries of what is possible, often discovering that the invisible—the intangible—holds the key to the visible world. Among these forces, expectation emerges as one of the most profound and misunderstood. It is more than mere belief or hope. Expectation, when fully understood, becomes a tool for actively shaping the reality we experience.

This book is a bold step into this uncharted territory. Building on the mathematical and philosophical foundations of BJ Klock's *Unified Field Theory (UFT)* and the principles of *Harmonic Thought*, this work seeks to define expectation not only as a psychological phenomenon but as a universal force with measurable, repeatable effects on the fabric of reality.

Our goal is to demonstrate that expectation is the bridge between thought and intention—the missing piece in our understanding of how the mind interacts with the cosmos. Through this framework, readers will learn to harness the power of expectation to align their beliefs with universal laws, creating a life of harmony, purpose, and fulfillment.

Connection to the Unified Field Theory (UFT)

The Unified Field Theory, as presented in BJ Klock's earlier works, unites the fundamental forces of the universe—gravity, electromagnetism, and the nuclear forces—into a single cohesive framework. Beyond its physical implications, the UFT bridges the divide between the scientific and metaphysical, offering a model for how thought, intention, and energy interact with the underlying structure of reality.

This book builds upon that foundation by introducing expectation as the next logical step in the evolution of this understanding. While thought initiates energy and intention directs it, expectation amplifies and aligns it with the harmonic principles of the universe. Expectation, we propose, is the mechanism through which human consciousness connects to the Ether and reshapes the probabilities of the physical world.

By integrating the principles of expectation into the UFT, this work expands our understanding of the cosmos and redefines the role of humanity as co-creators of reality.

What Readers Will Learn

1. **The Nature of Expectation:**

You will explore expectation as a force that operates beyond psychology, influencing the energetic and physical planes.

2. **A Mathematical Framework for Expectation:**

For the first time, we will present a model that mathematically describes how expectation interacts with intention, thought waves, and universal laws.

3. **Techniques to Master Expectation:**

Practical exercises and tools will show you how to align your thoughts and beliefs with harmonic principles to shape your reality intentionally.

4. **Applications in Personal and Societal Transformation:**

From individual empowerment to collective progress, you will learn how expectation can drive positive change on every level.

5. **The Future of Expectation and the UFT:**

Finally, you will see how expectation completes the triad of thought, intention, and reality, solidifying its place within the Unified Field Theory and humanity's evolving understanding of the cosmos.

A Call to the Reader

This book is more than an exploration of expectation; it is an invitation. It calls on each of us to question our beliefs, challenge our limitations, and harness the incredible power of our minds. Expectation is not just a passive state—it is an active force, a creative principle, and a universal law.

By the end of this journey, you will not only understand expectation; you will master it. Together, we will uncover how our expectations shape the world we live in and how aligning them with universal harmony can transform not only our lives but the future of humanity itself.

Let us begin.

Chapter 1: The Science of Expectation

Subtitle: *The Role of Anticipation in Human Experience*

1.1: The Cognitive Foundations of Expectation

Introduction
Expectation is woven into the fabric of human cognition. It is not merely a passive hope but an active mental process through which the brain predicts and prepares for outcomes. This ability has been essential to survival, influencing every decision, action, and reaction in our daily lives. But what if expectation is more than a cognitive phenomenon? What if it has measurable effects on the reality we experience?

To understand expectation as a universal force, we must first explore its cognitive origins and how the mind anticipates and interacts with the world.

The Predictive Mind
The human brain is a predictive engine, constantly analyzing past experiences to anticipate future events. This process occurs in the brain's **prefrontal cortex**, where information from memory, sensory input, and reasoning converge to create mental simulations of potential outcomes.

For instance, when you hear a sequence of musical notes, your brain instinctively predicts the next note based on patterns it has learned. This predictive mechanism operates in all aspects of life, from recognizing faces to navigating complex social interactions.

In the context of expectation, this predictive ability becomes a powerful tool. When aligned with focused intention, it primes the brain to recognize opportunities and solutions that might otherwise go unnoticed.

Neural Pathways and Habitual Expectation
Expectations are not random; they are shaped by neural pathways forged through repetition and experience. These pathways determine whether we expect success, failure, joy, or disappointment.

- **Positive Expectations:** Reinforce neural connections that support confidence, optimism, and action-oriented behavior.

- **Negative Expectations:** Strengthen pathways associated with fear, doubt, and inaction, creating a self-fulfilling cycle.

The plasticity of the brain—its ability to rewire itself—means that expectations can be consciously reshaped. This lays the foundation for aligning thought and intention with desired outcomes, a principle central to the framework introduced later in this book.

The Intersection of Mind and Reality
Modern neuroscience reveals that expectation influences not only mental processes but also physical and biological responses. For example:

- **The Placebo Effect:** Demonstrates that belief in a treatment can trigger real physiological changes, even when the treatment is inert.

- **Anticipatory Stress:** Shows how expecting a negative outcome can elevate cortisol levels, impacting the body's health and performance.

These findings suggest that expectation operates as a bridge between the mental and physical realms. By mastering this process, individuals can harness expectation as a tool for shaping their reality, both internally and externally.

Conclusion

Expectation begins as a cognitive process but extends far beyond the mind. It is deeply embedded in our biology, influencing everything from neural pathways to physical health. Understanding these foundations is the first step in uncovering how expectation interacts with universal laws to shape the world around us.

In the next section, we will explore how expectation evolved as a survival mechanism, highlighting its role as a driving force in human progress.

1.2: The Evolutionary Advantage of Expectation

Introduction
Throughout human history, the ability to anticipate and predict outcomes has been essential to survival. Expectation is not just a modern psychological phenomenon—it is a deeply ingrained evolutionary trait that has shaped the trajectory of our species. By understanding how expectation evolved as a survival mechanism, we can uncover its broader implications and its untapped potential to shape reality.

The Origins of Expectation in Survival
In the earliest stages of human evolution, expectation played a critical role in survival. Our ancestors relied on their ability to anticipate threats, predict weather patterns, and locate food sources. Those who could accurately expect outcomes based on environmental cues were better equipped to adapt, survive, and thrive.

- **Anticipating Danger:** Early humans learned to associate rustling in the bushes with the potential presence of predators. This expectation triggered immediate physiological responses—heightened senses, increased adrenaline, and readiness for action.
- **Predicting Resources:** Expectation enabled early societies to plan for seasonal changes, ensuring that food and shelter were available during harsh conditions.

These adaptive advantages became encoded in our neural architecture, making expectation a fundamental aspect of human cognition and behavior.

The Reward System and Expectation

Expectation is closely tied to the brain's reward system, particularly the release of dopamine, a neurotransmitter associated with pleasure and motivation.

- When an outcome matches or exceeds expectation, dopamine levels rise, reinforcing positive behavior.
- Conversely, when outcomes fall short of expectations, dopamine decreases, triggering a desire to adjust behavior or strategy.

This feedback loop drives learning and decision-making, ensuring that we adapt our actions to align with desired outcomes.

Example:

A hunter in ancient times anticipates the movement of prey based on previous experiences. When the hunt is successful, the brain releases dopamine, reinforcing the strategy. This same mechanism operates in modern contexts, such as setting and achieving personal or professional goals.

The Dual Nature of Expectation

While expectation has provided a survival advantage, it also comes with inherent risks. Misaligned or negative expectations can lead to self-sabotage or unnecessary stress.

- **Positive Expectations:** Drive motivation, innovation, and resilience, enabling individuals and societies to overcome challenges.
- **Negative Expectations:** Can result in fear-based decision-making, missed opportunities, and perpetuation of limiting beliefs.

The dual nature of expectation underscores the importance of mastering this force. By consciously aligning expectations with universal principles, individuals can harness its positive potential while minimizing its risks.

Expectation as a Driving Force in Human Progress
Beyond individual survival, expectation has fueled humanity's greatest achievements. From the invention of tools to the exploration of space, the ability to envision and anticipate possibilities has driven progress.

- **Scientific Discovery:** Every breakthrough begins with the expectation that new knowledge is attainable.
- **Social Movements:** Collective expectations of justice and equality have reshaped societies.
- **Technological Innovation:** The expectation of improvement has led to transformative advancements in communication, medicine, and energy.

As we refine our understanding of expectation, we unlock the potential to apply this force more intentionally—not only for personal growth but for societal transformation.

Conclusion
Expectation is more than a survival mechanism; it is the engine of human progress. By understanding its evolutionary origins and how it has shaped our development, we can begin to see expectation as a powerful tool for shaping the future.

In the next section, we will explore how expectation operates in psychology and behavior, providing real-world examples of its influence on individual and collective outcomes.

1.3: Expectation in Psychology and Behavior

Subtitle: How Belief Shapes Outcomes

Introduction
Expectation is deeply embedded in human psychology and behavior, influencing how we perceive, interpret, and react to the world. Whether conscious or subconscious, our expectations shape our actions and, in turn, the results we experience. From groundbreaking experiments in psychology to everyday experiences, the power of expectation reveals itself as a force that shapes reality.

This section will explore how expectation operates in the mind, its effects on behavior, and the profound impact it has on individual and collective experiences.

The Psychology of Expectation
Expectation is not just a passive belief—it is an active process that influences perception and decision-making. The brain filters incoming information to align with pre-existing expectations, often creating a self-fulfilling prophecy.

- **Selective Attention:** The brain prioritizes information that confirms our expectations while filtering out conflicting data.

Example: If you expect someone to dislike you, you are more likely to interpret neutral behavior as hostile, reinforcing your belief.

- **Confirmation Bias:** Once an expectation is formed, the mind seeks evidence to support it, even at the expense of contradictory facts.

These mechanisms show how expectation can shape not only perception but also the reality we create for ourselves.

The Placebo and Nocebo Effects
One of the most compelling demonstrations of the power of expectation is the placebo effect, where belief in a treatment produces real physiological changes. Conversely, the nocebo effect demonstrates how negative expectations can lead to adverse outcomes.

- **Placebo Effect:**

Example: A patient experiences pain relief after taking a sugar pill they believe is a powerful medication. Studies show that belief alone can activate the brain's pain-relief pathways.

- **Nocebo Effect:**

Example: A person experiences side effects after being told a harmless treatment might cause them, purely because of their expectation.

These phenomena highlight how expectation bridges the gap between mind and body, illustrating its potential to influence not just perception but tangible outcomes.

Expectation in Behavioral Psychology
Behavioral psychology provides further evidence of how expectation influences actions and outcomes.

- **The Pygmalion Effect:**

Teachers' high expectations of students often lead to improved performance, while low expectations can hinder achievement.
Key Insight: The expectations we project onto others can shape their reality.

- **Self-Fulfilling Prophecy:**

When individuals expect success or failure, their actions unconsciously align with those expectations, increasing the likelihood of the anticipated outcome.
Example: A job candidate confident in their ability is more likely to project enthusiasm and competence, increasing their chances of success.

These effects demonstrate that expectation is not just a passive state but an active force that directs behavior and influences outcomes.

The Ripple Effect of Collective Expectation
Expectation operates not only on an individual level but also within groups and societies.

- **Social Norms and Collective Beliefs:**

Shared expectations about acceptable behavior shape cultural norms and societal structures.
Example: The expectation of technological progress drives innovation, while collective fears about scarcity can lead to economic downturns.

- **Mass Expectation and Movements:**

Historical movements, such as the Civil Rights Movement or technological revolutions, were fueled by a collective expectation of change and progress.

When aligned, collective expectations can create powerful ripples that reshape societies, economies, and cultures.

Conclusion
The psychology of expectation reveals its profound impact on both individual behavior and collective experiences. Whether through the placebo effect, self-fulfilling prophecies, or societal movements, expectation proves to be a force that bridges belief and reality.

In the next chapter, we will take a deeper dive into how expectation interacts with the Ether, exploring its role as a universal substrate that harmonizes thought, intention, and reality.

Chapter 2: Expectation and the Ether

Subtitle: Resonance and the Universal Substrate

2.1: The Ether and Its Role in Reality

Introduction
The Ether has long been a topic of fascination and debate in both scientific and metaphysical circles. Once dismissed as an outdated concept, it has re-emerged in modern frameworks as a dynamic substrate that underlies all interactions in the universe. Thought, intention, and expectation do not exist in isolation; they ripple through this universal medium, connecting the mind to the cosmos.

In this section, we will explore the Ether as the bridge between expectation and reality, illustrating how it serves as the universal substrate through which energy and information flow.

What Is the Ether?
Historically, the Ether was imagined as a luminiferous medium, carrying light waves across the vacuum of space. While that specific notion was disproven, modern interpretations of the Ether align more closely with concepts like **quantum fields** or **vacuum energy**—a dynamic, invisible framework that underpins the universe.

The Ether can be understood as:

1. **A Carrier of Energy and Information:** It transmits waves, particles, and intentions across spacetime.

2. **A Substrate for Coherence:** It harmonizes interactions between physical and metaphysical phenomena, ensuring balance and connectivity.

3. **A Universal Memory:** The Ether stores patterns of energy, allowing resonance and feedback loops to emerge.

Expectation as a Ripple in the Ether
When we form an expectation, it sends a ripple through the Ether, similar to how a pebble disturbs the surface of a pond. These ripples carry energy and intention, interacting with the substrate to influence probabilities and outcomes.

- **Amplitude of Expectation:** Determines the strength of the ripple. Strong, focused expectations create larger ripples, increasing their influence on the surrounding substrate.

- **Frequency and Resonance:** Expectations aligned with universal harmonic principles, such as the Golden Ratio (ϕ), resonate more effectively, amplifying their impact.

Scientific Parallels: Quantum Fields and the Ether
Modern physics provides fascinating insights into how the Ether might operate on a quantum level.

- **Quantum Fields:** Every particle in the universe arises from an underlying quantum field, which governs its properties and interactions. Similarly, expectation interacts with the Ether as a field, shaping probabilities and outcomes.
  - *Example:* In quantum mechanics, the observer effect demonstrates how observation (or expectation) influences the behavior of particles.
- **Vacuum Energy:** The Ether can be likened to vacuum energy, which pervades all of spacetime and contains latent potential. Expectations may tap into this latent energy, directing it toward specific outcomes.

Resonance as the Key to Manifestation
The Ether responds most powerfully to expectations that are coherent and aligned with universal harmony. This concept of resonance is central to how expectation shapes reality:

1. **Alignment with ϕ:** The Golden Ratio governs proportions in nature, from galaxies to DNA. Expectations that harmonize with this principle find greater ease in manifesting outcomes.
2. **Coherence and Feedback:** Expectations that are clear, focused, and consistent create stronger, more stable ripples in the Ether.
3. **Universal Harmony:** The Ether acts as a regulator, ensuring that expectations aligned with balance and harmony integrate more seamlessly into reality.

Conclusion
The Ether is the universal substrate through which thought, intention, and expectation propagate. It connects the seen and unseen, providing a medium for shaping reality at both quantum and metaphysical levels. By understanding how expectation interacts with the Ether, we gain insight into the mechanics of manifestation and the principles of universal harmony.

In the next section, we will delve deeper into the quantum connection, examining how expectation aligns with phenomena like wave-particle duality and the observer effect.

2.2: The Quantum Connection

Subtitle: Expectation's Role in the Mechanics of Reality

Introduction

Quantum mechanics has long challenged our understanding of reality, revealing a world governed by probabilities, interconnectedness, and observer influence. In this realm, expectation finds a scientific parallel—a force that shapes not only perception but also outcomes.

This section explores how quantum phenomena, such as wave-particle duality and the observer effect, illustrate the interplay between expectation and the physical world. By connecting these principles to the Ether, we uncover a deeper understanding of how expectation acts as a bridge between thought and manifestation.

Wave-Particle Duality and the Role of Expectation

In quantum mechanics, particles such as electrons and photons can exist as waves or particles, depending on how they are observed. This phenomenon, known as **wave-particle duality**, challenges classical notions of fixed reality.

- **The Double-Slit Experiment:**

When particles are unobserved, they behave like waves, spreading across multiple paths simultaneously. However, when observed, they behave like particles, collapsing into a single location.

  - *Key Insight:* The act of observation—closely tied to expectation—determines the particle's behavior.

This demonstrates that reality is not fixed but influenced by the expectations and intentions of the observer.

The Observer Effect

The observer effect highlights how the presence and focus of an observer influence the behavior of quantum systems.

- **Connection to Expectation:**

The observer effect suggests that reality is not independent of consciousness. Instead, it responds to the observer's focus and anticipation. This aligns with the idea that expectation sends ripples through the Ether, directing energy toward specific outcomes.

  - *Example:* In experiments involving random number generators, participants' expectations have been shown to influence the sequence of generated numbers, further illustrating the tangible impact of focused thought.

Superposition and the Role of Anticipation

In quantum mechanics, particles exist in a state of **superposition**, where they simultaneously occupy multiple states until measured. This uncertainty mirrors the role of expectation in shaping reality:

- **Expectation as a Collapsing Force:**

Just as measurement collapses superposition into a specific state, expectation aligns probabilities within the Ether, collapsing them into manifested reality.

- *Example:* When we hold a strong expectation of success, we filter the vast array of potential outcomes into a single, desired result. This collapse reflects the interaction between expectation and the universal substrate.

Entanglement and Universal Connectivity

Quantum entanglement demonstrates that particles separated by vast distances remain connected, with changes to one particle instantly affecting the other.

- **Expectation and Entanglement:**

This phenomenon suggests that the universe operates as an interconnected whole. Expectation, as a ripple in the Ether, resonates across this interconnected field, influencing outcomes far beyond the immediate environment.

- *Example:* Collective expectations—such as societal optimism or fear—can ripple through populations, influencing global events and trends.

Expectation's Quantum Implications

By aligning expectation with quantum principles, we can better understand its role in shaping reality:

1. **Probability Alignment:** Expectation focuses energy, aligning probabilities in the Ether to favor desired outcomes.

2. **Observer Influence:** Consciousness, through expectation, acts as a creative force, interacting with the universe's foundational mechanics.

3. **Universal Resonance:** The interconnected nature of reality amplifies the influence of coherent expectations, spreading their effects across spacetime.

Conclusion

Quantum mechanics offers a scientific lens through which to view the power of expectation. From wave-particle duality to entanglement, these principles reveal a universe deeply influenced by observation, anticipation, and interconnectedness.

In the next section, we will explore how expectation aligns with universal patterns, particularly the Golden Ratio, to amplify its impact and manifest harmonious outcomes.

2.3: Aligning Expectation with Universal Patterns

Subtitle: The Role of Harmony in Amplifying Manifestation

Introduction
The universe operates according to profound patterns and principles that govern its structure and function. Among these, the Golden Ratio (ϕ) stands out as a universal constant that appears in everything from atomic particles to galaxies. By aligning expectation with these patterns, we can amplify its resonance, ensuring greater harmony and effectiveness in manifesting outcomes.

This section explores how expectation interacts with universal patterns, focusing on the Golden Ratio as a key amplifier of coherence and alignment in the Ether.

The Golden Ratio: Nature's Blueprint
The Golden Ratio, approximately 1.618, is a mathematical constant that appears throughout nature, art, and science. It represents a proportion that is both aesthetically pleasing and structurally efficient.

- **Examples in Nature:**
  - The spirals of galaxies and hurricanes.
  - The arrangement of leaves on a stem (phyllotaxis).
  - The proportions of the human body and DNA structure.

The ubiquity of the Golden Ratio suggests that it is not merely a mathematical curiosity but a fundamental principle of universal harmony.

Expectation and Resonance with ϕ
When an expectation aligns with universal patterns such as ϕ, it resonates more deeply within the Ether. This resonance amplifies its impact, creating greater coherence between thought, intention, and reality.

- **Harmonic Amplification:**

Just as a tuning fork amplifies sound when struck at its resonant frequency, expectations aligned with ϕ create ripples in the Ether that propagate more powerfully.

  - *Example:* Visualizing success in alignment with natural cycles (e.g., sunrise and growth seasons) can enhance the clarity and power of intention.

- **Coherence in Thought and Emotion:**

Expectations that are clear, focused, and emotionally charged align more effectively with the harmonic principles of the universe.

Patterns of Symmetry and Balance
Beyond the Golden Ratio, the universe exhibits symmetry and balance at all levels, from atomic structures to planetary orbits. Expectation interacts with these patterns by:

1. **Creating Symmetry:**

Expectations aligned with balance (e.g., focusing on both action and outcome) integrate seamlessly with the Ether.

2. **Restoring Harmony:**

When misaligned expectations create discord, re-centering them around universal patterns restores coherence.

- *Practical Insight:* If your expectations feel scattered or ineffective, aligning them with symmetrical or cyclical patterns can recalibrate their impact.

Practical Applications of Pattern Alignment
Harnessing the power of universal patterns requires intentional alignment. Here are some actionable steps:

1. **Visualization with ϕ:**

Incorporate visualizations that reflect the Golden Ratio, such as spirals or balanced proportions, to enhance the coherence of expectations.

2. **Using Natural Cycles:**

Set expectations in alignment with natural rhythms, such as lunar phases or seasonal changes, to amplify their resonance.

3. **Harmonic Thinking Exercises:**

Practice aligning thoughts and emotions with principles of symmetry and proportion, ensuring that expectation is clear, balanced, and purposeful.

Expectation as a Force of Harmony
When aligned with universal patterns, expectation becomes a force of harmony rather than disruption. It interacts with the Ether not as a chaotic ripple but as a purposeful wave, guiding energy and intention toward outcomes that reflect balance and beauty.

- **Case Study:**

In studies of collective intention, groups meditating on harmonious outcomes (e.g., peace or healing) have demonstrated measurable effects on societal conditions, suggesting that aligned expectations influence both individual and collective realities.

Conclusion
Expectation, when aligned with the Golden Ratio and other universal patterns, becomes a powerful tool for creating harmony in the physical and metaphysical realms. By understanding and applying these principles, we unlock the ability to manifest outcomes that are not only effective but deeply resonant with the fabric of reality.

In the next section, we will explore the role of coherence in expectation, examining how alignment and focus amplify its power within the Ether.

2.4: Coherence and Focus in Expectation

Subtitle: Amplifying Impact Through Alignment

Introduction
Expectation is most powerful when it is coherent and focused. Coherence ensures that the intention behind an expectation is aligned, consistent, and free of contradictory influences, while focus channels this expectation into a clear and purposeful direction. Together, these qualities amplify the impact of expectation, creating stronger ripples in the Ether and increasing the probability of manifestation.

In this section, we explore the mechanisms behind coherence and focus, their interplay, and practical strategies for enhancing these qualities in expectation.

Understanding Coherence in Expectation
Coherence refers to the alignment and harmony of thoughts, emotions, and intentions. When these elements are aligned, the energy behind an expectation becomes unified, magnifying its effect.

- **Internal Coherence:**

The alignment of thoughts, beliefs, and emotions within an individual.

  - *Example:* Expecting success while genuinely feeling confident and optimistic creates a coherent state.

- **External Coherence:**

The alignment of an individual's expectation with external conditions and universal principles.

- *Example:* Expecting personal growth while aligning actions with opportunities for learning and improvement.
- **The Role of ϕ:**

Coherence reflects the Golden Ratio's principle of balance, ensuring that expectations resonate harmoniously within the Ether.

The Power of Focus

Focus sharpens the energy of expectation, directing it toward a specific outcome. Just as a magnifying glass concentrates sunlight into a powerful beam, focused expectation channels mental and emotional energy into a singular purpose.

- **Clarity of Vision:**

Vague or scattered expectations dilute their impact. Focus requires a clear mental picture of the desired outcome.

  - *Example:* Instead of vaguely expecting "success," visualize a specific achievement, such as landing a new job or completing a creative project.

- **Sustained Attention:**

Focused expectation involves maintaining attention on the desired outcome over time, reinforcing its energy and resonance.

The Synergy Between Coherence and Focus

Coherence and focus are interdependent, working together to amplify expectation:

1. **Coherence Provides Stability:**

Ensures that the expectation is free of conflicting thoughts or emotions, creating a stable foundation.

2. **Focus Provides Direction:**

Channels the coherent energy toward a specific target, increasing its precision and effectiveness.

- *Example:* A coherent expectation of health (aligned thoughts, beliefs, and emotions) becomes more impactful when focused on a specific goal, such as recovering from an illness or improving fitness.

Practical Strategies for Enhancing Coherence and Focus

1. **Mindful Reflection:**

Regularly examine your thoughts and emotions to identify and resolve contradictions in your expectations.

- *Exercise:* Write down your expectation and note any doubts or fears that arise. Address these conflicts to achieve coherence.

2. **Visualization Techniques:**

Use detailed mental imagery to focus your expectation on a specific outcome.

- *Exercise:* Spend five minutes each day visualizing your desired result with as much clarity and emotion as possible.

3. **Emotional Alignment:**

Cultivate emotions that support your expectation, such as confidence, gratitude, and excitement.

- *Exercise:* Practice gratitude journaling to reinforce positive emotional states.

4. **Resonance with Universal Patterns:**

Align your expectation with harmonic principles, such as the Golden Ratio, to amplify its coherence.

- *Exercise:* Meditate on natural patterns, such as the rhythm of your breath or the symmetry of nature, to align your thoughts with universal harmony.

The Ripple Effect of Coherent and Focused Expectation

When expectation is coherent and focused, its impact extends far beyond the individual. It creates ripples in the Ether that influence probabilities, attract opportunities, and inspire action.

- *Case Study:*

Entrepreneurs who consistently align their expectations with their goals and maintain focus often report serendipitous opportunities, suggesting that coherent and focused intention interacts with external conditions to shape outcomes.

Conclusion

Coherence and focus are the twin pillars of powerful expectation. By aligning your thoughts, emotions, and intentions (coherence) and directing them with clarity and precision (focus), you can amplify the impact of your expectations, creating ripples in the Ether that align with your desired outcomes.

In the next chapter, we will delve into the mathematical framework of expectation, constructing equations that quantify its interaction with the Ether and its role in shaping reality.

Chapter 3: The Mathematical Framework of Expectation

Subtitle: Quantifying the Ripple Effect

3.1: The Expectation Wave Equation

Introduction
To fully harness the power of expectation, we must move beyond metaphors and conceptual understanding to a precise mathematical framework. By defining expectation as a wave-like phenomenon, we can begin to quantify its amplitude, frequency, coherence, and resonance.

In this section, we introduce the Expectation Wave Equation, a model that integrates expectation's energetic properties with the principles of the Ether and the Golden Ratio.

The Nature of Expectation as a Wave
Expectation behaves like a wave, propagating through the Ether as a ripple of energy and intention. This wave is defined by several key characteristics:

1. **Amplitude (A):** The strength or intensity of the expectation. Stronger emotions, beliefs, and focus increase amplitude.

2. **Frequency (f):** The rate at which the expectation cycles, reflecting its alignment with universal patterns.

3. **Phase Coherence (C):** The alignment of all components of the expectation—thoughts, emotions, and intentions. High coherence amplifies the wave's effectiveness.

4. **Resonance (R):** The degree to which the expectation aligns with the Golden Ratio (ϕ), enhancing its interaction with the Ether.

The Expectation Wave Equation
To quantify the influence of expectation, we define its wave as:

$$E(t) = A \cdot \sin(2 \pi f t + \phi) \cdot C \cdot R$$

Where:
- $E(t)$: The expectation at time t.
- A: Amplitude, representing the emotional and cognitive energy invested.
- f: Frequency, representing alignment with natural cycles or patterns.
- t: Time, representing the persistence and focus of the expectation.

- ϕ: Phase, representing alignment with harmonic principles (including the Golden Ratio).

- C: Coherence, a multiplier reflecting the internal alignment of thought and emotion (ranges from 0 to 1).

- R: Resonance, a multiplier reflecting alignment with universal patterns (ranges from 0 to 1).

Interpreting the Equation

1. **Amplitude (A) Matters:**

A high-amplitude expectation—driven by intense belief and emotional energy—creates a stronger ripple in the Ether.

2. **Frequency (f) Reflects Alignment:**

Expectations aligned with universal patterns (e.g., natural rhythms, ϕ) resonate more effectively, increasing their impact.

3. **Coherence (C) Magnifies Impact:**

Coherent expectations—free of internal contradictions—propagate more powerfully than scattered or conflicting ones.

4. **Resonance (R) Amplifies Effectiveness:**

Expectations aligned with the Golden Ratio (ϕ) achieve harmonic resonance, ensuring maximum interaction with the Ether.

Practical Applications

1. **Visualizing Expectation Waves:**

   - Visualize your expectation as a wave propagating through the Ether. Ensure the amplitude (intensity), coherence, and resonance are aligned with your goal.

2. **Measuring Coherence and Resonance:**

   - Use self-reflection to assess coherence (are your thoughts and emotions aligned?) and resonance (is your expectation in harmony with natural patterns?).

3. **Iterative Refinement:**

   - Adjust the amplitude, coherence, and resonance of your expectation over time to strengthen its ripple effect.

Conclusion

The Expectation Wave Equation provides a quantitative framework for understanding and amplifying the power of expectation. By aligning amplitude, frequency, coherence, and resonance, individuals can create ripples in the Ether that align with their desired outcomes.

In the next section, we will expand this framework to explore how collective expectation interacts with the Ether, amplifying its impact on societal and global scales.

3.2: Coherence and Intention in Collective Expectation

Subtitle: Amplifying Impact Through Unified Thought

Introduction
While individual expectation creates ripples in the Ether, the combined power of collective expectation can generate waves strong enough to shape societies and influence global events. When groups align their thoughts, emotions, and intentions, the resulting coherence amplifies their collective influence.

This section explores how collective expectation operates, the principles that enhance its coherence, and the societal transformations it can achieve when aligned with universal patterns.

The Nature of Collective Expectation
Collective expectation arises when a group of individuals shares a unified vision or belief. Its effectiveness depends on several key factors:

1. **Shared Intention:** A common goal or purpose that aligns individual expectations into a cohesive force.

2. **Group Coherence:** The degree of alignment in the group's thoughts, emotions, and actions. Higher coherence amplifies the ripple effect in the Ether.

3. **Scale:** The size of the group contributing to the expectation. Larger groups generate more significant waves, but coherence remains critical.

The Mathematics of Collective Expectation

The power of collective expectation can be represented as the summation of individual expectation waves, taking coherence into account:

$$E_{\text{collective}}(t) = \sum_{i=1}^n E_i(t) \cdot C_{\text{group}}$$

Where:

- $E_{\text{collective}}(t)$: The collective expectation at time t.
- $E_i(t)$: The expectation of each individual i.
- n: The number of individuals in the group.
- C_{group}: The group's overall coherence (ranges from 0 to 1).

Enhancing Group Coherence

1. **Shared Vision:**

   • Establishing a clear and compelling goal that resonates with all participants ensures alignment of intention.

   • *Example:* A community meditating on world peace focuses their collective expectation on a shared outcome.

2. **Emotional Synchronization:**

   • Encouraging positive, shared emotions such as hope, gratitude, and confidence amplifies coherence.

   • *Example:* Large-scale events like concerts or protests often generate emotional resonance, enhancing collective expectation.

3. **Physical and Energetic Synchronization:**

   • Group activities such as synchronized breathing, chanting, or meditation create physical and energetic alignment.

   • *Example:* Studies on group meditation have demonstrated measurable reductions in local crime rates, suggesting a ripple effect from coherent collective intention.

The Golden Ratio in Collective Expectation
Collective expectation achieves maximum resonance when it aligns with universal patterns, including the Golden Ratio (ϕ):

- **Harmonic Leadership:** Leaders who embody principles of balance and harmony inspire greater coherence within their groups.

- **Proportional Participation:** Groups that balance diverse perspectives while maintaining a unified purpose mirror the harmony of ϕ, amplifying their collective influence.

Real-World Examples of Collective Expectation

1. **Historical Movements:**

    - The Civil Rights Movement and similar efforts succeeded in part due to the coherence and shared vision of their participants.

2. **Scientific Collaboration:**

    - Large-scale research projects, such as the Human Genome Project, demonstrate how collective expectation drives innovation and discovery.

3. **Cultural Phenomena:**

    - Viral trends and global movements, fueled by collective belief and participation, highlight the power of aligned thought and emotion.

Potential and Responsibility of Collective Expectation
The power of collective expectation comes with immense potential and responsibility. While it can drive positive change, misaligned or fear-driven expectations can create discord and unintended consequences.

- **Positive Impact:** Groups aligned with universal principles can inspire societal transformation, addressing challenges such as inequality, climate change, and conflict.

- **Avoiding Discord:** Ensuring that collective expectation is guided by coherent, positive intentions minimizes the risk of amplifying fear or division.

Conclusion
Collective expectation is a potent force that amplifies the power of individual intention through alignment and coherence. By understanding and applying its principles, we can harness this energy to create meaningful, harmonious change on a societal and global scale.

In the next section, we will examine how these principles can be applied to practical systems, such as governance, education, and innovation, to shape a better future.

3.3: Practical Systems for Collective Expectation

Subtitle: Designing Frameworks to Harness Unified Thought

Introduction
Collective expectation has the power to transform societies, innovate systems, and address global challenges. However, harnessing this potential requires structured,

scalable frameworks that channel unified thought into actionable systems. By embedding principles of coherence, alignment, and resonance into governance, education, and technological innovation, we can design systems that amplify collective expectation for the greater good.

In this section, we explore practical methods to integrate collective expectation into societal structures, fostering collaboration and harmony on a global scale.

Governance Models Built on Collective Coherence
Governance systems that align with universal principles foster trust, collaboration, and progress by harnessing the power of collective expectation.

1. **Participatory Leadership:**

    - Involves citizens actively shaping policies and decisions through structured forums of collective thought.

    - *Example:* Citizen assemblies or deliberative democracies provide platforms for shared visions to emerge, ensuring alignment with the public's expectations.

2. **Harmonic Policies:**

    - Policies crafted with principles of balance, fairness, and proportionality resonate with collective well-being.

    - *Example:* Implementing economic models that balance growth with sustainability, reflecting the harmony of the Golden Ratio.

3. **Expectation-Driven Metrics:**

    - Governance systems can adopt metrics to assess and align policies with public expectations and universal principles.

    - *Example:* Measuring societal coherence through indices that track collective satisfaction, equity, and harmony.

Education for Unified Thought
An education system designed to foster collective coherence equips individuals with the tools to align their expectations with universal principles, amplifying societal harmony.

1. **Teaching Coherence and Focus:**

    - Curricula that include meditation, visualization, and reflective practices help students align their thoughts and emotions.

- *Example:* Mindfulness programs in schools have demonstrated improvements in focus, emotional regulation, and academic performance.

2. **Emphasizing Collaborative Learning:**

- Encouraging group projects and discussions cultivates shared vision and collective expectation.

- *Example:* Problem-based learning where students collaborate to address real-world challenges fosters collective creativity.

3. **Incorporating Universal Principles:**

- Lessons on natural patterns, such as the Golden Ratio, can inspire students to align their intentions with universal harmony.

Technological Innovation for Collective Expectation

Advancements in technology offer unprecedented opportunities to channel collective expectation into systems that enhance global coherence.

1. **Platforms for Unified Thought:**

- Digital platforms can facilitate collective intention by enabling millions to share and focus on a common goal.

- *Example:* Global meditation apps that synchronize users to focus on world peace or environmental restoration.

2. **AI-Driven Insights:**

- Artificial intelligence can analyze patterns in collective expectation, providing actionable insights for governance, education, and societal planning.

- *Example:* AI tools that identify emerging global priorities based on collective sentiment and behavior.

3. **Resonant Technologies:**

- Devices and systems designed to amplify harmonic frequencies can support aligned collective intention.

- *Example:* Energy grids that integrate with natural cycles to enhance sustainability and efficiency.

Case Studies in Collective Expectation Systems

1. **Global Movements:**

- The Paris Agreement on climate change demonstrates the power of aligned global expectations to address pressing challenges.

    2. **Educational Transformation:**

    - Finland's collaborative education model, which emphasizes equity and creativity, aligns with principles of coherence and shared intention.

    3. **Technological Breakthroughs:**

    - Open-source platforms like Wikipedia illustrate how collective expectation can create valuable, universally accessible resources.

Challenges and Opportunities

While collective expectation holds immense potential, its implementation requires addressing key challenges:

    1. **Overcoming Mistrust:**

    - Building systems that foster transparency and inclusivity ensures trust and alignment within groups.

    2. **Managing Discordant Expectations:**

    - Balancing competing interests and expectations requires mediation strategies rooted in universal principles.

    3. **Scaling Coherence:**

    - Creating scalable frameworks for large populations demands innovative approaches that maintain alignment and focus.

Conclusion

By embedding principles of collective expectation into governance, education, and technology, we can design systems that harmonize human potential with universal principles. These frameworks serve as catalysts for societal progress, fostering unity and amplifying the ripple effect of aligned intention.

In the next chapter, we will explore the ethical considerations and responsibilities associated with harnessing collective expectation, ensuring that its power is used for the greater good.

Chapter 4: Ethical Considerations in Harnessing Expectation

Subtitle: Aligning Power with Responsibility

4.1: The Ethical Dimensions of Expectation

Introduction

Harnessing the power of expectation, whether individually or collectively, comes with profound ethical responsibilities. As expectation shapes reality through ripples in the Ether, its misuse can lead to unintended consequences, discord, or harm. This chapter examines the ethical dimensions of expectation, emphasizing the need to align its application with universal principles of harmony and equity.

The Dual Nature of Expectation

Expectation, like any force, can be used constructively or destructively. Its dual nature necessitates careful consideration of intentions and outcomes.

1. **Constructive Use:**
   - Focusing on outcomes that promote well-being, progress, and harmony.
   - *Example:* Collective meditations for peace or innovation that benefits humanity.

2. **Destructive Use:**
   - Misaligned expectations driven by fear, greed, or domination create ripples that disrupt harmony.
   - *Example:* Fear-driven narratives that propagate anxiety and division.

Guiding Principles for Ethical Expectation

1. **Alignment with Universal Harmony:**
   - Ensure expectations resonate with principles of balance and the Golden Ratio, fostering constructive outcomes.

2. **Consideration of Ripple Effects:**
   - Recognize that expectation influences not only the immediate environment but also distant and interconnected systems.
   - *Example:* Policies or technologies inspired by collective expectations should prioritize long-term sustainability.

3. **Equity and Inclusivity:**
   - Ethical expectation accounts for diverse perspectives and strives to benefit the collective rather than a select few.

Case Study: Historical Examples of Ethical and Unethical Use

1. **Ethical Collective Expectation:**

    - The Civil Rights Movement demonstrated aligned collective expectation for justice and equality, fostering societal transformation.

2. **Unethical Manipulation of Expectation:**

    - Propaganda campaigns in authoritarian regimes leveraged collective expectation to manipulate and suppress, creating discord.

Accountability in Expectation

1. **Personal Responsibility:**

    - Individuals must reflect on their intentions and consider the broader impact of their expectations.

    - *Example:* Entrepreneurs envisioning innovations must evaluate both societal benefits and potential unintended consequences.

2. **Collective Responsibility:**

    - Groups and leaders hold a heightened responsibility to align expectations with ethical frameworks.

    - *Example:* Corporations leveraging collective expectation through marketing must avoid fostering unrealistic or harmful aspirations.

Tools for Ethical Alignment

1. **Reflective Practices:**

    - Regular introspection to ensure expectations are aligned with ethical principles.

2. **Impact Assessment:**

    - Evaluating the potential ripple effects of individual and collective expectations before taking action.

3. **Guidance from Universal Principles:**

    - Using principles such as the Golden Ratio to assess the harmony and balance of intended outcomes.

Conclusion

The ethical dimensions of expectation demand mindfulness, accountability, and alignment with universal principles. By ensuring that expectation is used constructively and inclusively, we can harness its transformative power responsibly.

4.2: Ethical Challenges in Collective Expectation

Subtitle: Navigating Complexity and Responsibility

Introduction
Harnessing the power of collective expectation amplifies its impact but also introduces significant ethical challenges. The collective nature of expectation creates complex dynamics, including competing interests, unintended consequences, and the risk of manipulation. This section explores these challenges and provides strategies for navigating them while maintaining alignment with universal principles.

Key Ethical Challenges

1. **Competing Expectations:**

    - When multiple groups with differing expectations interact, discord can arise, disrupting harmony in the Ether.

    - *Example:* Conflicting political ideologies can create societal polarization.

2. **Manipulation of Expectation:**

    - Powerful entities, such as corporations or governments, may exploit collective expectation to serve self-interests rather than the greater good.

    - *Example:* Fear-driven campaigns that influence public behavior for profit or control.

3. **Unintended Consequences:**

    - Even well-intentioned collective expectations can lead to outcomes that are misaligned with ethical principles.

    - *Example:* Rapid technological innovation spurred by collective demand may inadvertently harm the environment.

4. **Scale and Accountability:**

    - The larger the group contributing to collective expectation, the more challenging it becomes to ensure alignment and accountability.

Strategies for Addressing Ethical Challenges

1. **Facilitating Coherence Across Groups:**

    • Create platforms for dialogue and collaboration to align competing expectations toward common goals.

    • *Example:* International summits addressing global challenges like climate change foster collective coherence.

2. **Transparency and Oversight:**

    • Ensure that leaders and organizations guiding collective expectation operate transparently and are held accountable.

    • *Example:* Publicly disclosing the motivations and goals behind large-scale initiatives to build trust and alignment.

3. **Ethical Education:**

    • Teach individuals and groups to critically evaluate collective expectations and their potential consequences.

    • *Example:* Programs that emphasize mindfulness, critical thinking, and alignment with universal principles.

4. **Alignment with Universal Harmony:**

    • Use principles like the Golden Ratio and balance to evaluate whether collective expectations resonate harmoniously with the greater good.

Case Studies of Ethical Challenges

1. **The Environmental Movement:**

    • While aligned with the universal principle of sustainability, collective expectations have faced challenges from competing economic interests and misinformation campaigns.

2. **The Rise of Social Media Movements:**

    • Platforms that amplify collective expectation have demonstrated both positive outcomes (e.g., global awareness of human rights issues) and ethical concerns (e.g., the spread of misinformation and polarization).

3. **Technological Advancements:**

    • Collective demand for innovation has driven rapid progress but often at the expense of ethical considerations, such as data privacy and environmental impact.

The Role of Leadership in Ethical Collective Expectation
Ethical leadership plays a pivotal role in addressing challenges and aligning collective expectation with universal principles.

1. **Inspiring Harmony:**

    - Leaders should embody balance, inclusivity, and vision, fostering coherence within their groups.

    - *Example:* Visionary leaders like Mahatma Gandhi aligned collective expectation with ethical principles of nonviolence and equity.

2. **Prioritizing the Greater Good:**

    - Decisions should be guided by long-term benefits for humanity and the planet, rather than short-term gains.

    - *Example:* Policies that prioritize sustainability and equity over profit-driven motives.

3. **Engaging with Diverse Perspectives:**

    - Ethical leaders actively listen to and integrate diverse viewpoints to ensure inclusivity and balance.

Conclusion
Navigating the ethical challenges of collective expectation requires awareness, collaboration, and a steadfast commitment to universal principles. By fostering coherence, ensuring accountability, and aligning with harmony, we can address these challenges responsibly and unleash the transformative potential of collective expectation.

In the next section, we will explore how to design systems and frameworks that institutionalize ethical principles within collective expectation.

4.3: Institutionalizing Ethics in Collective Systems

Subtitle: Embedding Universal Principles into Societal Frameworks

Introduction
For collective expectation to serve as a force for progress and harmony, ethical principles must be institutionalized within societal systems. This involves creating governance, education, and technological frameworks that align with universal principles such as balance, inclusivity, and sustainability. By embedding these

values, we can ensure that collective expectation fosters equitable and constructive outcomes.

This section explores the design and implementation of ethical systems that amplify the positive potential of collective expectation while minimizing risks and unintended consequences.

Principles for Ethical System Design

1. **Alignment with Universal Harmony:**

    - Systems should reflect the principles of balance, proportionality, and resonance found in the Golden Ratio (ϕ).

    - *Example:* A governance model that balances economic growth with environmental sustainability mirrors the harmony of natural patterns.

2. **Transparency and Accountability:**

    - Ethical systems require open processes and mechanisms for evaluating outcomes to build trust and maintain coherence.

    - *Example:* Regular audits and public reporting on the alignment of policies with ethical goals.

3. **Inclusivity and Diversity:**

    - Incorporating diverse perspectives ensures that collective expectation aligns with the greater good rather than the interests of a few.

    - *Example:* Educational curricula that represent global cultures and viewpoints to foster inclusivity.

Governance Frameworks for Ethical Collective Expectation

1. **Harmonic Decision-Making Models:**

    - Implement decision-making processes that prioritize balance and fairness.

    - *Example:* Weighted voting systems that consider the proportional impact of decisions on different stakeholders.

2. **Dynamic Feedback Loops:**

    - Incorporate mechanisms that adapt policies based on evolving collective expectations and outcomes.

- *Example:* Environmental policies that adjust dynamically based on real-time ecological data.

3. **Ethical Oversight Councils:**

- Establish independent bodies to ensure that collective expectation initiatives align with universal principles.

- *Example:* Councils that evaluate the ethical implications of new technologies or societal initiatives.

Educational Systems for Ethical Alignment

1. **Teaching Universal Principles:**

- Incorporate lessons on balance, coherence, and harmony into educational systems to foster aligned thought from an early age.

- *Example:* Courses that explore the interconnectedness of humanity and the environment.

2. **Encouraging Collaborative Learning:**

- Design group-based projects and problem-solving activities to teach the value of collective intention and shared goals.

- *Example:* Students collaborating on sustainability initiatives as part of their curriculum.

3. **Ethics and Empathy Training:**

- Programs that cultivate empathy and ethical reasoning prepare individuals to contribute positively to collective expectation.

- *Example:* Role-playing scenarios to practice resolving conflicts and aligning intentions ethically.

Technological Systems for Ethical Application

1. **AI and Machine Learning for Ethical Insights:**

- Develop AI systems that analyze collective expectation trends and provide guidance aligned with universal principles.

- *Example:* AI-driven tools that evaluate the sustainability of policy decisions.

2. **Expectation-Driven Platforms:**

- Create digital platforms that facilitate collective coherence by synchronizing user intentions around shared goals.

- *Example:* Apps that coordinate global meditations for environmental healing or conflict resolution.

3. **Resonant Design Principles:**

- Ensure that technology integrates seamlessly with natural systems, promoting harmony and sustainability.

- *Example:* Energy systems that align with planetary cycles to minimize waste and disruption.

Challenges in Institutionalizing Ethics

1. **Overcoming Resistance to Change:**

- Implementing ethical systems may face resistance from entrenched interests or conflicting priorities.

2. **Balancing Complexity and Simplicity:**

- Designing systems that are both robust and accessible requires careful consideration of scalability and user engagement.

3. **Ensuring Global Collaboration:**

- Aligning diverse cultures, nations, and organizations with shared ethical principles requires fostering mutual understanding and trust.

Case Studies in Ethical System Design

1. **The United Nations Sustainable Development Goals (SDGs):**

- A global framework aligned with principles of inclusivity and sustainability, fostering collective coherence on issues such as poverty, education, and climate change.

2. **Open-Source Movements:**

- Platforms like Linux and Wikipedia demonstrate the potential of collective intention and ethical collaboration in creating shared resources.

3. **Community-Led Initiatives:**

- Local governance models that prioritize equity and transparency, such as participatory budgeting programs, illustrate the power of ethical alignment at smaller scales.

Conclusion

Institutionalizing ethics in collective systems ensures that the power of collective expectation is harnessed constructively and responsibly. By embedding principles of balance, inclusivity, and accountability into governance, education, and technology, we can create a foundation for sustained harmony and progress.

In the next section, we will explore the potential for global collaboration to further amplify the impact of ethical collective expectation on a planetary scale.

4.4: Global Collaboration and Ethical Unity

Subtitle: Aligning Humanity Through Collective Purpose

Introduction
In an interconnected world, the potential of collective expectation extends beyond individual nations or communities—it holds the power to unify humanity around shared goals and principles. Global collaboration rooted in ethical alignment is essential for addressing existential challenges, such as climate change, inequality, and technological disruption.

This section examines how collective expectation can serve as the foundation for global collaboration, emphasizing shared purpose, mutual trust, and universal principles.

The Need for Global Collaboration

1. **Interconnected Challenges:**

   • Issues like environmental sustainability, economic inequality, and global health are interconnected, requiring collective action to achieve meaningful progress.

2. **Shared Responsibility:**

   • The Ether connects all of humanity, emphasizing our shared responsibility to align intentions with the greater good.

3. **Leveraging Diversity:**

   • Global collaboration benefits from the diverse perspectives, skills, and cultural wisdom of people across the world.

Principles for Ethical Global Collaboration

1. **Common Vision:**

    • Establish a unifying goal that transcends individual interests and inspires collective action.

    • *Example:* Achieving carbon neutrality through international cooperation and innovation.

2. **Mutual Respect and Equity:**

    • Ensure all voices are valued and represented, fostering inclusivity and trust.

    • *Example:* Collaborative decision-making processes that give equal weight to both developed and developing nations.

3. **Alignment with Universal Principles:**

    • Use the Golden Ratio and balance as guiding frameworks for creating policies and solutions.

    • *Example:* International agreements that balance economic development with environmental sustainability.

Frameworks for Global Collaboration

1. **Ethical Alliances:**

    • Establish global coalitions committed to ethical principles, such as fairness, transparency, and sustainability.

    • *Example:* Expanding the United Nations to include an "Ethical Council" that evaluates the long-term impact of international decisions.

2. **Technological Platforms:**

    • Utilize digital tools to synchronize global expectations and facilitate collaboration across borders.

    • *Example:* A decentralized platform for tracking progress on shared goals, powered by blockchain technology for transparency.

3. **Global Education Initiatives:**

    • Develop programs that teach universal principles and foster a sense of interconnectedness among future generations.

    • *Example:* International exchange programs focused on sustainability and innovation.

The Role of Leadership in Global Collaboration

1. **Visionary Leadership:**

    - Leaders must articulate a compelling vision that inspires unity and action.

    - *Example:* Speeches or campaigns that emphasize humanity's shared destiny and interconnectedness.

2. **Facilitators of Trust:**

    - Building trust among nations, organizations, and individuals is essential for fostering collaboration.

    - *Example:* Hosting transparent dialogues that address past grievances and build mutual understanding.

3. **Ethical Decision-Making:**

    - Leaders should prioritize decisions that align with universal harmony, ensuring long-term benefits over short-term gains.

Examples of Successful Global Collaboration

1. **The Paris Climate Agreement:**

    - Demonstrates the power of collective expectation in addressing global challenges through shared commitment to sustainability.

2. **The Human Genome Project:**

    - A global scientific collaboration that aligned diverse expertise to achieve groundbreaking progress in genetics.

3. **The International Space Station (ISS):**

    - A symbol of global unity, showcasing how collective expectation can drive innovation and exploration.

Challenges to Global Collaboration

1. **Cultural and Political Differences:**

    - Misaligned priorities and conflicting ideologies can hinder collective action.

2. **Mistrust Among Stakeholders:**

    - Historical grievances and power imbalances may create barriers to collaboration.

3. **Scaling Coherence:**

    • Ensuring alignment and coherence across diverse groups requires continuous effort and innovative approaches.

Strategies to Overcome Challenges

1. **Bridge-Building Initiatives:**

    • Develop programs that foster cultural understanding and empathy among diverse groups.

    • *Example:* Global youth summits that bring together leaders of tomorrow to discuss shared goals.

2. **Transparent Systems:**

    • Create mechanisms for monitoring progress and holding stakeholders accountable.

    • *Example:* A public dashboard that tracks global initiatives and highlights areas for improvement.

3. **Focus on Shared Humanity:**

    • Emphasize our commonalities rather than differences, fostering a sense of unity and purpose.

Conclusion
Global collaboration, rooted in ethical unity and guided by collective expectation, has the potential to transform humanity's trajectory. By aligning efforts with universal principles, we can address pressing challenges and create a future defined by harmony, equity, and shared purpose.

In the next chapter, we will explore how to validate the principles of expectation through measurable experimentation and observation, bridging the gap between theoretical understanding and practical application.

Chapter 5: Validating the Principles of Expectation

Subtitle: Bridging Theory and Reality

5.1: Observable Effects of Expectation

Introduction

The power of expectation has been theorized as a force capable of shaping reality through the Ether, aligning with universal principles. To elevate this concept from theoretical to empirical, it is essential to identify observable effects and measure their impact. This chapter explores real-world phenomena that demonstrate how expectation influences outcomes, from personal experience to large-scale societal shifts.

Real-World Phenomena Demonstrating Expectation

1. **The Placebo Effect:**

    - One of the most well-documented examples of expectation influencing reality, where belief in a treatment often leads to measurable improvements.

    - *Example:* Studies show patients taking sugar pills experience pain relief comparable to actual medication when they expect to feel better.

2. **Self-Fulfilling Prophecies:**

    - Expectations about a person, situation, or outcome often lead to behaviors that bring those expectations to fruition.

    - *Example:* Teachers' higher expectations for students correlate with better academic performance, known as the "Pygmalion Effect."

3. **The Observer Effect in Quantum Mechanics:**

    - Observations or measurements at the quantum level influence the behavior of particles, suggesting a connection between expectation and physical outcomes.

    - *Example:* The double-slit experiment demonstrates how observation alters wave-particle duality.

4. **Collective Movements and Cultural Shifts:**

    - Shared expectations within societies often lead to significant cultural or political transformations.

    - *Example:* The civil rights movement in the United States was propelled by collective belief in equality and justice, reshaping societal norms.

Key Variables That Influence Observable Effects

1. **Coherence:**

    - The alignment of thoughts and emotions strengthens the influence of expectation on reality.

- *Example:* Group meditations have been linked to measurable decreases in local crime rates, as coherence amplifies collective intention.

2. **Repetition:**

- Repeatedly focusing on an expectation reinforces its influence on the Ether.
- *Example:* Daily affirmations and visualizations practiced over time lead to noticeable changes in behavior and outcomes.

3. **Emotional Intensity:**

- Strong emotions tied to an expectation create a more significant impact.
- *Example:* Hope and confidence often enhance the likelihood of success, while fear and doubt can hinder progress.

Potential Metrics for Measuring Expectation's Effects

1. **Physiological Changes:**

- Heart rate variability and brainwave coherence can be measured to track the alignment of thought and emotion.

2. **Behavioral Outcomes:**

- Analyzing patterns of success or change correlated with specific expectations.
- *Example:* Comparing goal-setting practices with achieved outcomes across diverse groups.

3. **Environmental Impact:**

- Examining shifts in physical systems, such as local ecosystems or atmospheric conditions, in response to collective expectation initiatives.

Challenges in Observing and Measuring Expectation

1. **Subjectivity:**

- Expectations are inherently personal, making them difficult to quantify uniformly.

2. **External Influences:**

- Environmental and societal factors can complicate the isolation of expectation's direct effects.

3. **Replication and Consistency:**

- Reproducing outcomes across different individuals or groups requires rigorous methodologies.

Case Studies of Observable Effects

1. **Water Crystallization Experiments:**

   - Studies suggest that focused thought and intention influence the structure of water crystals, highlighting a potential link between expectation and physical matter.

2. **Random Number Generator Studies:**

   - Experiments where participants influence random outcomes through focused intention demonstrate measurable deviations in expected randomness.

3. **Global Meditation Events:**

   - Large-scale meditations correlated with reductions in conflict or natural disasters suggest the power of collective coherence.

Conclusion
Observable effects of expectation provide compelling evidence of its influence on reality. While challenges remain in measurement and replication, the existing phenomena offer a foundation for further exploration. In the next section, we will propose specific experiments designed to validate the principles of expectation rigorously and systematically.

5.2: Proposed Experiments for Validating Expectation

Subtitle: Designing Tests to Bridge Theory and Empirical Evidence

Introduction
To establish expectation as a measurable force that influences reality, rigorous experimentation is required. This section outlines proposed studies designed to validate the principles of expectation, focusing on replicability, measurable outcomes, and alignment with universal principles. These experiments aim to test the impact of individual and collective expectation on physical systems, human behavior, and societal outcomes.

1. Testing Individual Expectation

Experiment A: Influence on Random Number Generators (RNGs)

- **Objective:** Assess whether focused thought can influence the output of RNGs.

- **Methodology:**

    1. Participants focus on achieving specific patterns or outcomes in the generated numbers.

    2. Data from control (no intention applied) and experimental groups are compared.

    3. Statistical analysis determines whether deviations from randomness are significant.

- **Expected Outcome:** Higher coherence and emotional engagement correlate with deviations in randomness aligned with participants' expectations.

Experiment B: Intention and Water Crystallization

- **Objective:** Test whether focused intention affects the structure of water crystals.

- **Methodology:**

    1. Participants direct positive or negative intentions toward separate water samples.

    2. Water is frozen, and the resulting crystal structures are analyzed for symmetry and complexity.

    3. Control samples are left undisturbed.

- **Expected Outcome:** Water samples exposed to positive intention display more harmonious and intricate crystal structures.

2. Testing Collective Expectation

Experiment C: Group Meditation and Environmental Impact

- **Objective:** Examine whether group coherence influences local environments, such as crime rates or atmospheric conditions.

- **Methodology:**

    1. Groups participate in synchronized meditations focusing on specific outcomes (e.g., reduced crime or rainfall).

    2. Compare environmental or social metrics during meditation periods with baseline data.

    3. Analyze data across multiple locations to ensure replicability.

- **Expected Outcome:** Significant changes in target metrics during meditation periods.

Experiment D: Collective Visualization and Economic Outcomes

- **Objective:** Test whether collective expectation can influence economic indicators, such as stock market trends or consumer behavior.

- **Methodology:**

    1. Large groups visualize specific economic scenarios (e.g., growth in specific industries).

    2. Compare the outcomes with control periods where no collective visualization occurs.

    3. Use sentiment analysis to track changes in public perception and behavior.

- **Expected Outcome:** Correlation between collective visualization and targeted economic trends.

3. Testing Expectation in Human Behavior

Experiment E: The Pygmalion Effect in Education

- **Objective:** Investigate how teachers' expectations influence student performance.

- **Methodology:**

    1. Randomly assign students to two groups. Teachers are informed that one group has higher potential (regardless of actual ability).

    2. Track academic performance over a semester.

    3. Compare outcomes between the two groups.

- **Expected Outcome:** Students perceived as high-potential show greater improvement, validating the impact of expectation on behavior.

Experiment F: Expectation in Athletic Performance

- **Objective:** Assess whether visualization and expectation improve physical performance in athletes.

- **Methodology:**

1. Divide athletes into control and experimental groups.

2. The experimental group practices guided visualization and expectation exercises before events.

3. Compare performance metrics between the groups.

- **Expected Outcome:** Athletes using visualization outperform the control group.

4. Cross-Disciplinary Validation

Experiment G: Thought-Wave Measurement in Neural Activity

- **Objective:** Detect changes in brainwave coherence during expectation-focused exercises.
- **Methodology:**

1. Use EEG devices to monitor participants during focused intention exercises.

2. Compare neural activity during expectation-focused periods with baseline activity.

3. Correlate changes in coherence with outcomes in related physical or behavioral tests.

- **Expected Outcome:** High coherence in brainwave activity correlates with improved outcomes in intention-based tasks.

Experiment H: Collective Sentiment and Weather Patterns

- **Objective:** Investigate whether collective expectation influences localized weather phenomena.
- **Methodology:**

1. Organize large-scale synchronized visualizations targeting specific weather outcomes (e.g., rain in drought areas).

2. Compare weather data during visualization periods with historical trends and control periods.

3. Include multiple global locations for replicability.

- **Expected Outcome:** Observable deviations in weather patterns aligned with collective visualization.

Challenges and Ethical Considerations

1. **Replicability:**

   • Ensuring consistency across studies requires robust methodologies and controlled environments.

2. **Bias Mitigation:**

   • Addressing potential biases in participants and researchers ensures objective results.

3. **Ethical Use of Findings:**

   • Avoiding misuse of expectation-based techniques (e.g., manipulation or exploitation) is essential.

Conclusion

These proposed experiments aim to validate the influence of expectation on reality, providing empirical evidence to bridge the gap between theory and practice. By designing studies that encompass individual, collective, and interdisciplinary approaches, we can further illuminate the role of expectation as a universal force.

5.3: The Role of Coherence in Measurable Outcomes

Subtitle: Amplifying the Impact of Expectation Through Alignment

Introduction

Coherence is the alignment of thoughts, emotions, and intentions within an individual or group, creating a unified and harmonious state. In the context of expectation, coherence acts as a multiplier, amplifying the effects of intention and expectation on reality. This section explores the mechanisms through which coherence enhances measurable outcomes, its role in individual and collective practices, and practical ways to cultivate it.

The Science of Coherence

1. **Neural Coherence:**

   • Coherence in the brain occurs when neural activity aligns, creating harmonious wave patterns.

   • *Example:* EEG studies show that meditative states associated with coherent brainwaves enhance cognitive performance and emotional regulation.

2. **Emotional Coherence:**

   • Alignment of emotional states reduces internal conflict, enabling clearer and more focused expectations.

   • *Example:* Heart rate variability (HRV) studies reveal that coherence between the heart and brain improves physiological and psychological resilience.

3. **Group Coherence:**

   • Collective coherence emerges when individuals align their intentions and emotions, creating a synergistic effect.

   • *Example:* Studies on synchronized group activities, such as prayer or meditation, show measurable impacts on social and environmental conditions.

Mechanisms of Coherence in Amplifying Outcomes

1. **Focused Energy:**

   • Coherence reduces the dispersion of mental and emotional energy, concentrating it toward a single goal.

   • *Example:* Athletes visualizing their performance with coherent focus often achieve better results.

2. **Resonance with the Ether:**

   • Aligned intentions resonate more effectively within the Ether, creating stronger ripples that influence reality.

   • *Example:* Collective visualization exercises have been linked to measurable reductions in local crime rates during synchronization periods.

3. **Minimizing Interference:**

   • Coherence eliminates internal contradictions and competing expectations, reducing noise in the energetic signal.

   • *Example:* Individuals practicing coherent visualization report fewer obstacles in achieving their goals.

Cultivating Coherence

1. **Individual Practices:**

   • Techniques to align thoughts, emotions, and intentions:

- *Meditation:* Focus on a single intention while maintaining emotional calm.
- *Visualization:* Create vivid mental images of desired outcomes while sustaining emotional resonance.
- *Breathwork:* Practice rhythmic breathing to synchronize the mind and body.

2. **Group Practices:**
- Methods for fostering coherence in groups:
  - *Guided Meditations:* Facilitators lead groups in synchronized focus on a shared goal.
  - *Rituals and Ceremonies:* Structured activities that promote alignment through symbolic actions.
  - *Resonant Environments:* Use of music, visuals, or physical spaces designed to evoke harmony.

3. **Technological Aids:**
- Tools to measure and enhance coherence:
  - *Biofeedback Devices:* Measure HRV and brainwave patterns to provide real-time feedback.
  - *Coherence Apps:* Facilitate synchronized group activities through virtual platforms.
  - *Wearable Technology:* Devices that track physiological coherence and encourage mindfulness.

Applications of Coherence in Real-World Scenarios

1. **Personal Growth:**
- Coherence enhances goal achievement, resilience, and well-being.
- *Example:* Entrepreneurs using coherence practices report increased clarity and success in their ventures.

2. **Social Movements:**
- Collective coherence drives effective activism and societal change.

- *Example:* Peaceful demonstrations grounded in collective intention often achieve greater impact than fragmented movements.

3. **Environmental Healing:**

- Group coherence amplifies environmental initiatives, fostering balance and regeneration.
- *Example:* Coordinated global meditations for ecological restoration have been associated with localized improvements in environmental health.

Challenges in Achieving Coherence

1. **Internal Distractions:**

- Mental and emotional turbulence disrupts alignment, reducing coherence.

2. **Group Dynamics:**

- Misaligned goals or interpersonal conflicts hinder collective coherence.

3. **Skepticism and Resistance:**

- Lack of belief in the power of coherence can limit participation and commitment.

Conclusion
Coherence is the foundation for amplifying the measurable outcomes of expectation. By aligning thoughts, emotions, and intentions, individuals and groups can create focused, powerful ripples within the Ether, influencing reality in profound ways. The next section will explore how coherence translates expectation into actionable, real-world changes.

5.4: Connecting Thought to Action

Subtitle: Bridging Expectation and Tangible Outcomes

Introduction
Expectation becomes a transformative force when it bridges the gap between mental intention and physical reality. This section explores the pathways through which coherent thought translates into meaningful action and measurable results. By understanding how expectation shapes decision-making, behavior, and outcomes, individuals and groups can harness its power for personal growth, societal progress, and technological innovation.

The Pathway from Thought to Action

1. **Mental Alignment:**

    • Coherent thoughts create a clear blueprint in the mind, reducing ambiguity and uncertainty.

    • *Example:* A student who envisions their success with clarity is more likely to take steps that align with that vision.

2. **Emotional Resonance:**

    • Strong emotional connection to an expectation motivates action and sustains focus.

    • *Example:* Entrepreneurs driven by passion for their vision demonstrate higher resilience and creativity in overcoming obstacles.

3. **Behavioral Patterns:**

    • Repeated focus on an expectation reinforces habits and behaviors that align with desired outcomes.

    • *Example:* Athletes who visualize their performance regularly develop training routines that reflect their mental image of success.

4. **External Manifestation:**

    • Thought waves interact with the Ether, influencing circumstances and opportunities that align with expectations.

    • *Example:* Collective expectations for social change often create the conditions for policy shifts or cultural transformation.

Key Factors in Translating Thought to Action

1. **Clarity of Intention:**

    • A well-defined expectation ensures focused energy and reduces competing thoughts.

    • *Example:* Creating a clear, actionable plan increases the likelihood of achieving a goal.

2. **Alignment with Universal Principles:**

    • Expectations that align with balance, harmony, and sustainability are more likely to manifest effectively.

    • *Example:* A company prioritizing ethical practices often experiences long-term growth and stability.

3. **Consistent Reinforcement:**

   • Repetition strengthens the connection between expectation and action.

   • *Example:* Daily affirmations and visualizations create a mental framework that influences daily decisions.

Practical Techniques for Bridging Thought and Action

1. **Visualization Practices:**

   • Create vivid mental images of desired outcomes, incorporating sensory details to enhance emotional resonance.

   • *Example:* Visualizing a successful presentation improves confidence and preparation.

2. **Goal Setting with Intention:**

   • Combine specific, measurable goals with a coherent focus on the desired outcome.

   • *Example:* Setting milestones for a fitness journey while visualizing the end result.

3. **Behavioral Alignment:**

   • Ensure daily actions align with expectations to create a feedback loop of progress.

   • *Example:* Writing down small, actionable steps toward a larger goal reinforces commitment and coherence.

4. **Group Coherence Activities:**

   • Synchronize efforts with others to amplify collective impact.

   • *Example:* Teams aligning on shared objectives during meetings experience higher productivity and success rates.

Applications in Real-World Scenarios

1. **Personal Development:**

   • Harnessing expectation for self-improvement:

      • *Example:* Using thought-action alignment to overcome procrastination and achieve career goals.

2. **Societal Change:**

- Aligning collective expectations to influence policies and cultural norms:
  - *Example:* Grassroots movements leveraging collective intention to drive legislative reform.

3. **Technological Innovation:**
   - Applying thought-action principles in research and development:
     - *Example:* Collaborative innovation labs using group coherence to accelerate breakthroughs in clean energy.

4. **Healing and Wellness:**
   - Translating expectation into actionable health practices:
     - *Example:* Patients combining visualization with medical treatments to enhance recovery outcomes.

Challenges in Bridging Thought to Action

1. **Lack of Focus:**
   - Diffuse or conflicting expectations hinder progress.

2. **Emotional Disconnection:**
   - Expectations lacking emotional resonance fail to inspire meaningful action.

3. **Resistance to Change:**
   - Habits and environmental factors may counteract efforts to align actions with expectations.

The Ripple Effect of Thought into Action

- **Individual Impact:**
  - A single aligned individual can inspire change in their immediate environment, creating a ripple effect.
  - *Example:* A teacher implementing thought-action principles influences students, who in turn shape their communities.

- **Collective Influence:**
  - Group coherence amplifies individual actions, creating larger waves of transformation.

- *Example:* Collective movements for sustainability inspire global adoption of renewable energy practices.

Conclusion
Connecting thought to action is the ultimate expression of coherent expectation. By aligning intention with behavior, individuals and groups can transform abstract goals into tangible realities. In the next chapter, we will explore how to apply these principles in personal transformation, empowering readers to master their thoughts and actions for meaningful change.

Chapter 6: Personal Transformation

Subtitle: Mastering Your Thoughts to Master Your Reality

6.1: Aligning Expectation with Universal Principles
Subtitle: Practical Techniques for Achieving Coherence

Introduction
Personal transformation begins with the alignment of thoughts and expectations with universal principles such as balance, harmony, and resonance. This alignment is not merely philosophical but practical, creating a foundation for meaningful and lasting change. In this section, we explore actionable techniques to achieve coherence and align expectations with the natural laws that govern reality.

Why Alignment Matters

1. **Resonance with Universal Forces:**

   - Alignment ensures that thoughts and actions harmonize with the Ether, amplifying their impact.

   - *Example:* Intentions grounded in harmony often lead to smoother execution and fewer obstacles.

2. **Clarity and Focus:**

   - Aligned expectations reduce internal contradictions, creating a clear path toward desired outcomes.

   - *Example:* Defining a singular, coherent goal minimizes distractions and confusion.

3. **Empowerment Through Understanding:**

   - Aligning with universal principles fosters a sense of purpose and connection to the greater whole.

- *Example:* Feeling connected to a broader mission sustains motivation during challenging periods.

Techniques for Aligning Expectation

1. **Visualization Exercises:**

    - Craft mental images that reflect your desired outcomes, ensuring they align with principles of harmony and balance.

    - *Practice:* Spend 10 minutes daily visualizing your goals, focusing on the sensory details and emotional resonance of success.

2. **Journaling for Clarity:**

    - Write down your expectations, goals, and the principles they align with to reinforce coherence.

    - *Practice:* Begin each journal entry by identifying how your goal contributes to balance or harmony in your life.

3. **Breathwork for Emotional Coherence:**

    - Use rhythmic breathing to align emotional and mental states, creating a foundation for clear and focused expectation.

    - *Practice:* Inhale for a count of four, hold for four, exhale for four, and repeat for five minutes.

4. **Daily Affirmations:**

    - Repeat statements that reinforce aligned expectations, embedding them in your subconscious.

    - *Practice:* Create affirmations that include universal principles, such as, "I align my actions with harmony and balance, achieving success with ease."

The Role of Reflection in Alignment

1. **Self-Assessment:**

    - Regularly evaluate whether your thoughts and actions reflect your core principles.

    - *Exercise:* At the end of each day, ask yourself: "Did my actions today align with my expectations and universal principles?"

2. **Learning from Misalignment:**

- Identify instances of misalignment as opportunities for growth and adjustment.
- *Example:* Reflect on decisions driven by fear or impatience and reframe them within a harmonious perspective.

3. **Celebrating Alignment:**
- Acknowledge moments where your expectations and actions harmonized with universal forces, reinforcing positive behavior.
- *Practice:* Keep a log of achievements that reflect aligned expectations to track progress over time.

Case Studies of Alignment in Action

1. **Personal Achievement:**
- *Example:* A professional seeking career advancement visualized their desired role daily, aligning their actions with principles of persistence and collaboration. Within months, they secured their dream position.

2. **Health and Wellness:**
- *Example:* An individual recovering from illness used breathwork and affirmations to align their expectations with resilience, experiencing accelerated healing.

3. **Creative Pursuits:**
- *Example:* An artist aligned their expectations with the principle of balance, creating a body of work that resonated deeply with audiences and achieved critical acclaim.

Challenges to Achieving Alignment

1. **Overcoming Self-Doubt:**
- Negative thought patterns disrupt coherence, creating misalignment.
- *Solution:* Practice affirmations and mindfulness to counteract self-doubt.

2. **Balancing Competing Goals:**
- Conflicting priorities dilute focus and hinder alignment.
- *Solution:* Prioritize goals that align most closely with universal principles.

3. **Sustaining Alignment Over Time:**

- Maintaining focus requires consistent effort and reflection.
- *Solution:* Establish daily routines that reinforce aligned expectations.

Conclusion
Aligning expectation with universal principles is the cornerstone of personal transformation. By cultivating coherence and clarity through practical techniques, individuals can harness the power of expectation to create meaningful, sustainable change. The next section will delve into the importance of focused intention as a tool for amplifying clarity and direction in achieving aligned goals.

6.2: Focused Intention

Subtitle: Harnessing the Power of Clarity and Direction

Introduction
Focused intention is the art of channeling your mental, emotional, and physical energies toward a singular goal. It refines expectation into a precise and directed force, amplifying its power to influence reality. This section explores the mechanisms, benefits, and techniques of focused intention, offering practical methods to transform abstract desires into actionable outcomes.

The Science of Focused Intention

1. **Cognitive Clarity:**
   - When attention is focused, the brain filters out distractions, enhancing problem-solving and decision-making abilities.
   - *Example:* Neuroscientific studies show that heightened focus activates the prefrontal cortex, the center for goal-directed behavior.

2. **Energetic Precision:**
   - Intention focused with clarity and emotional resonance sends a coherent signal into the Ether, increasing its impact.
   - *Example:* Athletes who visualize specific aspects of their performance achieve greater accuracy and success.

3. **Reinforcing Neural Pathways:**
   - Repeated focus on a goal strengthens neural pathways, embedding the desired outcome into subconscious behavior.

- *Example:* Entrepreneurs who practice focused intention report heightened creativity and resilience in their pursuits.

Why Focused Intention Matters

1. **Eliminates Ambiguity:**

   - Vague expectations dilute mental energy, while focused intention sharpens clarity and direction.

   - *Example:* A writer with a clear vision for their book is more likely to finish it efficiently than one with scattered ideas.

2. **Increases Emotional Resonance:**

   - Focused intention amplifies the emotional connection to a goal, sustaining motivation and commitment.

   - *Example:* Visualizing the joy of achieving a fitness goal strengthens the resolve to maintain consistent effort.

3. **Creates a Feedback Loop:**

   - Focused intention aligns actions with desired outcomes, reinforcing positive behaviors and results.

   - *Example:* A student focusing on academic success is more likely to study effectively, resulting in better grades and reinforcing their commitment.

Techniques for Cultivating Focused Intention

1. **Single-Point Meditation:**

   - Focus on a single image, word, or sensation to train the mind in concentrated attention.

   - *Practice:* Spend 10 minutes visualizing a detailed image of your goal while maintaining deep, rhythmic breathing.

2. **Intentional Journaling:**

   - Write specific goals and the steps needed to achieve them, refining your focus daily.

   - *Practice:* Begin each day by outlining your primary intention and reflecting on its alignment with your values.

3. **Affirmation and Visualization Cycles:**

- Combine affirmations with vivid mental imagery to reinforce focused intention.
- *Practice:* Pair affirmations like "I am focused, determined, and aligned with my goal" with visualizations of achieving success.

4. **Creating Vision Boards:**

- Use physical or digital boards to map out your intentions with images, quotes, and milestones.
- *Practice:* Update your board regularly to reflect evolving goals and achievements.

The Role of Coherence in Focused Intention

1. **Mental Alignment:**

- Coherence ensures that all thoughts and emotions support a unified intention, reducing internal conflict.
- *Example:* A scientist working on a breakthrough aligns their focus with curiosity, discipline, and passion, avoiding distractions.

2. **Group Focused Intention:**

- Collaborative intention amplifies the impact of shared goals.
- *Example:* Teams practicing group visualization report improved synergy and productivity.

3. **Resonance with the Ether:**

- Focused and coherent intentions create stronger ripples within the universal field.
- *Example:* Communities focusing on sustainable goals often see accelerated progress toward ecological restoration.

Overcoming Challenges to Focused Intention

1. **Distracted Attention:**

- In today's fast-paced world, distractions dilute focus.
- *Solution:* Establish regular practices such as meditation or journaling to retrain attention.

2. **Emotional Inconsistency:**

- Fluctuations in emotional state disrupt coherence.
- *Solution:* Use techniques like breathwork and mindfulness to stabilize emotions before focusing intention.

3. **Competing Priorities:**
- Conflicting goals scatter energy and hinder progress.
- *Solution:* Prioritize intentions that align with universal principles and long-term vision.

Applications of Focused Intention

1. **Personal Achievement:**
- *Example:* An artist uses focused intention to refine their creative vision, resulting in a critically acclaimed masterpiece.

2. **Wellness and Healing:**
- *Example:* Patients practicing focused intention alongside medical treatment report faster recovery and improved well-being.

3. **Social Innovation:**
- *Example:* Leaders using focused intention to align their teams achieve breakthroughs in technology, governance, or education.

Conclusion
Focused intention transforms expectation into a laser-like force that bridges thought and action. By cultivating clarity, coherence, and emotional resonance, individuals and groups can channel their energy toward meaningful and measurable outcomes. In the next section, we will explore how harmonic thought principles can be applied to healing and resilience, empowering readers to align their minds and bodies with universal balance.

6.3: Healing and Resilience

Subtitle: Aligning Thought and Expectation to Empower the Mind and Body

Introduction
Healing and resilience are deeply influenced by the mind's ability to align thought and expectation with the body's natural capacity for balance and regeneration. By applying the principles of coherence and focused intention, individuals can enhance their physical, emotional, and mental well-being. This section explores how

harmonic thought fosters healing and strengthens resilience, offering techniques and examples for practical application.

The Science of Healing and Resilience

1. **The Mind-Body Connection:**

   - Research demonstrates that thoughts and emotions influence physiological processes.

   - *Example:* Positive expectations about recovery have been shown to improve outcomes in medical treatments, a phenomenon known as the placebo effect.

2. **Stress and Coherence:**

   - Chronic stress disrupts coherence between the mind and body, impairing immune function and emotional resilience.

   - *Example:* Practices like mindfulness and meditation reduce cortisol levels, promoting harmony and health.

3. **Resilience Through Neural Plasticity:**

   - The brain's ability to adapt and rewire itself—neuroplasticity—enables individuals to cultivate resilience in the face of adversity.

   - *Example:* Survivors of trauma often rebuild emotional strength through intentional thought practices.

The Role of Expectation in Healing

1. **Positive Expectation:**

   - Belief in a positive outcome creates physiological changes that accelerate healing.

   - *Example:* Patients who visualize their recovery report reduced pain and faster healing times.

2. **The Power of Coherence:**

   - Aligned thoughts and emotions amplify the body's natural regenerative processes.

   - *Example:* Guided imagery sessions have been linked to improved immune responses in patients undergoing cancer treatment.

3. **The Ripple Effect in Resilience:**

- Focused thought not only promotes healing but also fortifies resilience, creating a feedback loop of well-being.
- *Example:* Athletes recovering from injury use visualization to regain strength and confidence, enabling faster return to peak performance.

Techniques for Healing and Resilience

1. **Visualization for Healing:**
   - Envision the body repairing itself, focusing on the desired outcome with clarity and emotional resonance.
   - *Practice:* Spend 10 minutes daily visualizing cells regenerating or pain dissipating, accompanied by slow, rhythmic breathing.

2. **Affirmations for Resilience:**
   - Repeat affirmations that reinforce strength, adaptability, and recovery.
   - *Practice:* Use statements like, "My body is strong, my mind is calm, and I am healing with each breath."

3. **Coherence Practices:**
   - Synchronize the mind and body through activities that promote coherence, such as yoga or biofeedback exercises.
   - *Practice:* Engage in heart-centered breathing to align emotional states with physiological harmony.

4. **Guided Meditation for Recovery:**
   - Follow meditations designed to foster relaxation and focused healing.
   - *Practice:* Use recorded or live sessions that guide you through visualizing light or energy restoring balance to your body.

Applications of Healing and Resilience

1. **Chronic Illness Management:**
   - Patients with chronic conditions use harmonic thought to improve quality of life.
   - *Example:* A patient with autoimmune disease visualizes harmonious cellular interactions, reporting decreased symptoms over time.

2. **Trauma Recovery:**

- Individuals recovering from emotional or physical trauma apply coherent thought to rebuild resilience.
- *Example:* A trauma survivor uses affirmations and mindfulness to reframe their experience, fostering emotional growth.

3. **Everyday Stress Reduction:**

- Applying these principles in daily life reduces stress and prevents burnout.
- *Example:* Professionals practicing daily visualization and coherence exercises report enhanced productivity and emotional well-being.

Challenges in Healing and Resilience

1. **Overcoming Negative Thought Patterns:**

- Chronic negativity impedes the body's ability to heal.
- *Solution:* Reframe negative thoughts into constructive affirmations through mindfulness practices.

2. **Skepticism Toward Non-Traditional Methods:**

- Resistance to applying mental techniques for physical healing may hinder their adoption.
- *Solution:* Highlight evidence-based studies and real-world examples to build confidence.

3. **Maintaining Consistency:**

- Healing and resilience require ongoing effort and commitment.
- *Solution:* Integrate practices into daily routines to ensure regularity and sustainability.

Conclusion

Healing and resilience are not just physical processes; they are deeply intertwined with the mind's ability to align thought, expectation, and intention. By leveraging harmonic thought techniques, individuals can unlock their innate capacity for regeneration and adaptability. The next section will explore how achieving balance through harmonic thought enhances overall well-being and aligns personal goals with universal harmony.

6.4: Achieving Balance

Subtitle: Aligning Personal Goals with Universal Harmony

Introduction

Balance is the cornerstone of a harmonious life. Achieving balance involves aligning personal goals and actions with the universal principles of coherence, resonance, and harmony. This section explores how individuals can cultivate balance by integrating harmonic thought into daily life, fostering a sense of purpose and connection with the greater whole.

The Importance of Balance

1. **Foundation for Well-Being:**

   - Balance between mental, emotional, and physical states enhances overall health and happiness.

   - *Example:* Individuals who maintain work-life balance report higher levels of satisfaction and productivity.

2. **Alignment with Universal Principles:**

   - Actions aligned with universal laws of harmony and coherence reduce resistance and create smoother pathways to success.

   - *Example:* Entrepreneurs who prioritize ethical practices often experience long-term stability and growth.

3. **Sustainability and Resilience:**

   - Balance fosters adaptability, enabling individuals to navigate challenges without losing alignment with their goals.

   - *Example:* Leaders who integrate balance into their decision-making inspire trust and collaboration among their teams.

Steps to Achieve Balance

1. **Define Core Values:**

   - Identify personal principles that align with universal harmony.

   - *Practice:* Reflect on questions like, "What do I value most in life?" and "How can I align my goals with those values?"

2. **Set Intentional Goals:**

   - Establish goals that reflect a balance between ambition and sustainability.

- *Practice:* Create SMART goals (Specific, Measurable, Achievable, Relevant, Time-bound) rooted in universal principles.

3. **Create a Daily Routine:**

- Build habits that support mental, emotional, and physical alignment.

- *Practice:* Incorporate practices like meditation, journaling, and physical exercise into your routine to maintain balance.

4. **Cultivate Mindful Decision-Making:**

- Evaluate choices based on their alignment with long-term goals and universal harmony.

- *Practice:* Before making decisions, ask yourself, "Does this action support balance in my life and the lives of others?"

Practical Techniques for Maintaining Balance

1. **Energy Management:**

- Balance energy expenditure across work, relationships, and personal pursuits to avoid burnout.

- *Practice:* Use time-blocking techniques to allocate focused periods for different aspects of life.

2. **Harmonic Thought Reflection:**

- Regularly assess whether your thoughts and actions align with the principles of harmonic thought.

- *Practice:* Dedicate 5 minutes daily to reflecting on how your day supported or disrupted balance.

3. **Adaptability Through Recalibration:**

- Adjust goals and actions as circumstances change to maintain alignment.

- *Practice:* Review and update your goals monthly to ensure they remain relevant and achievable.

4. **Seek External Feedback:**

- Engage trusted mentors or peers to provide perspective on your progress.

- *Practice:* Schedule regular check-ins with accountability partners to discuss your alignment and balance.

Applications of Achieving Balance

1. **Personal Growth:**

    • *Example:* A student balances academic and social pursuits, achieving both high performance and a fulfilling personal life.

2. **Professional Success:**

    • *Example:* A manager balances assertiveness and empathy, fostering a productive and supportive work environment.

3. **Community Engagement:**

    • *Example:* Community leaders align local initiatives with principles of sustainability, creating long-term impact and resilience.

4. **Environmental Stewardship:**

    • *Example:* Individuals who balance consumption with conservation contribute to ecological harmony and sustainability.

Challenges in Achieving Balance

1. **Competing Priorities:**

    • Managing multiple demands can create imbalances.

    • *Solution:* Prioritize activities that align most closely with universal principles and long-term goals.

2. **Overcommitment:**

    • Taking on too many responsibilities disrupts balance.

    • *Solution:* Learn to say no and delegate tasks when possible.

3. **Resistance to Change:**

    • Habitual patterns may hinder efforts to achieve balance.

    • *Solution:* Use gradual adjustments to shift habits toward alignment with harmonic thought.

Conclusion
Achieving balance is an ongoing process that requires self-awareness, intentionality, and adaptability. By aligning personal goals with universal harmony, individuals can foster a life of fulfillment, resilience, and purpose. In the next chapter, we will

transition from personal transformation to societal transformation, exploring how harmonic thought can serve as a blueprint for collective progress.

Chapter 7: Expectation in the Unified Field Theory

Subtitle: A New Force in the Universal Framework

7.1: Completing the Triad
Subtitle: Thought, Intention, and Expectation as Interconnected Universal Forces

Introduction
Within the framework of the Unified Field Theory (UFT), thought, intention, and expectation form a triad of universal forces that shape reality. Each element complements the others: thought initiates, intention directs, and expectation amplifies. Together, they create a cohesive framework for understanding how human agency influences the physical and metaphysical realms.

The Interconnected Roles of Thought, Intention, and Expectation

1. **Thought as the Blueprint:**

    - Thought provides the initial structure for potential outcomes.

    - *Example:* An architect visualizing a building creates the mental foundation for its eventual construction.

2. **Intention as the Driver:**

    - Intention focuses energy and aligns action with the envisioned outcome.

    - *Example:* The architect's deliberate planning and resource allocation turn the blueprint into actionable steps.

3. **Expectation as the Amplifier:**

    - Expectation bridges the gap between potential and manifestation, reinforcing coherence and alignment.

    - *Example:* The architect's belief in the project's success inspires confidence in collaborators, accelerating progress.

The Synergy of the Triad in Action

1. **Harmonic Resonance:**

    - Thought, intention, and expectation, when aligned, create a resonance that influences the Ether, amplifying their impact on reality.

- *Example:* A musician composing a symphony uses thought to conceptualize, intention to arrange, and expectation to perform, creating a masterpiece that moves audiences.

2. **Coherence Across Scales:**

- The triad functions at both micro and macro levels, shaping individual experiences and collective realities.
- *Example:* A grassroots movement aligns shared thought, collective intention, and unified expectation to achieve systemic change.

3. **Feedback Loops:**

- Expectation enhances the clarity of thought and the precision of intention, creating a positive feedback loop.
- *Example:* A scientist expecting to validate a hypothesis refines their thought process and experimental design, increasing the likelihood of discovery.

The Role of Expectation in the Universal Framework

1. **Predictive Resonance:**

- Expectation acts as a predictive force, aligning possibilities with probabilities.
- *Example:* Athletes visualizing victory align their performance with the expectation of success, increasing their chances of winning.

2. **Expectation as a Universal Law:**

- Like gravity or electromagnetism, expectation operates as a force that influences both matter and energy.
- *Example:* Collective expectations of societal progress create tangible shifts in technology, governance, and culture.

Conclusion

The triad of thought, intention, and expectation reveals the interconnected nature of human agency within the UFT. By recognizing expectation as an amplifying force, we gain a deeper understanding of how individuals and societies shape reality. In the next section, we will explore the predictive nature of expectation and its interaction with the Ether, further solidifying its role in the universal framework.

7.2: Expectation as Predictive Resonance

Subtitle: Aligning Possibilities with Probabilities

Introduction

Expectation serves as a bridge between potential and reality, a predictive force that aligns possibilities with probabilities. It operates within the Unified Field Theory (UFT) as a dynamic influence, shaping outcomes by harmonizing thought and intention with the fabric of the universe. This section explores the mechanisms of expectation as predictive resonance and its profound implications for personal and collective experiences.

Understanding Predictive Resonance

1. **Expectation as a Resonant Frequency:**

   - Each expectation carries a vibrational signature that resonates with specific outcomes in the Ether.

   - *Example:* A confident speaker expecting their message to inspire an audience often sees that expectation fulfilled, as their resonance aligns with the audience's response.

2. **Probabilities Within the Field:**

   - The Ether holds infinite possibilities, and expectation filters these into probable outcomes.

   - *Example:* A scientist expecting success in an experiment unconsciously adjusts variables and actions to align with the desired result, increasing its probability.

3. **The Amplifying Role of Emotion:**

   - Emotions attached to expectation strengthen its resonance, making outcomes more likely.

   - *Example:* A student expecting to succeed in an exam feels motivated and focused, amplifying their ability to study effectively and perform well.

Mechanics of Expectation in the UFT

1. **Energy Flow and Coherence:**

   - Expectation directs energy flow within the Ether, creating a coherent pathway for manifestation.

   - *Example:* An entrepreneur expecting a successful launch naturally attracts collaborators and resources aligned with their vision.

2. **Alignment and Probabilistic Collapse:**

- Expectation acts as a magnet, pulling potentialities into alignment with a singular outcome.
- *Example:* A sports team collectively expecting to win creates a unified energy field that supports cohesive teamwork and optimal performance.

3. **Feedback Loops in Reality Formation:**

- Expectation reinforces itself through feedback loops, solidifying desired outcomes into reality.
- *Example:* A patient expecting to recover experiences positive feedback from each small improvement, accelerating their healing process.

Practical Applications of Predictive Resonance

1. **Goal Setting with Expectation:**

- Combine clear, focused goals with unwavering expectation to increase success.
- *Practice:* Visualize the outcome vividly while maintaining emotional resonance with the desired result.

2. **Group Dynamics and Collective Expectation:**

- Teams or communities sharing a unified expectation can manifest extraordinary results.
- *Example:* A nonprofit organization expecting to make a significant impact aligns its efforts and attracts like-minded donors and volunteers.

3. **Expectation in Education and Learning:**

- Teachers with high expectations for students often inspire greater achievement.
- *Example:* Educators who believe in their students' potential encourage them to meet those expectations, resulting in higher performance.

4. **Harnessing Expectation in Health and Wellness:**

- Patients expecting positive outcomes respond better to treatments, creating measurable improvements.
- *Example:* Medical studies show that patients with optimistic outlooks recover faster from surgeries or illnesses.

Challenges in Harnessing Predictive Resonance

1. **Overcoming Negative Expectations:**

   • Negative expectations can create self-fulfilling prophecies, reinforcing undesirable outcomes.

   • *Solution:* Reframe limiting beliefs into empowering ones through affirmations and focused intention.

2. **Managing Unrealistic Expectations:**

   • Unrealistic or poorly aligned expectations disrupt coherence and hinder manifestation.

   • *Solution:* Ground expectations in achievable goals while maintaining optimism and flexibility.

3. **Skepticism and Resistance:**

   • Doubts about the effectiveness of expectation can weaken its resonance.

   • *Solution:* Use small, incremental successes to build confidence in the process.

Case Studies and Real-World Examples

1. **The Power of Collective Expectation:**

   • *Example:* Historical movements such as civil rights advancements were driven by collective expectation for justice and equality.

2. **Expectation in Leadership:**

   • *Example:* Leaders who expect innovation and success inspire their teams to rise to the challenge, often exceeding expectations.

3. **Scientific Validation:**

   • *Example:* Experiments in random number generation influenced by focused thought demonstrate the measurable impact of expectation on physical systems.

Conclusion

Expectation as predictive resonance is a profound force within the Unified Field Theory. By aligning thought, intention, and expectation, individuals and groups can harmonize with the probabilities of their desired outcomes, shaping reality with precision and purpose. In the next section, we will explore the interaction between expectation and the Ether, delving deeper into how this universal force influences the physical and metaphysical realms.

7.3: The Ether and Expectation

Subtitle: How Expectation Interacts with the Universal Medium

Introduction
The Ether, as defined within the Unified Field Theory (UFT), is the universal substrate through which all forces, matter, and energy interact. Expectation, as a resonant force, acts as a dynamic agent within the Ether, influencing its structure and behavior to align potentialities with desired outcomes. This section explores the interaction between expectation and the Ether, demonstrating how this connection bridges the metaphysical and physical realms.

The Role of the Ether in Manifestation

1. **Ether as a Conduit:**

   - The Ether facilitates the flow of energy and information, serving as the medium through which expectation operates.

   - *Example:* Just as sound waves travel through air, expectation creates vibrational patterns in the Ether that resonate with specific outcomes.

2. **Field Dynamics and Coherence:**

   - Expectation introduces coherence into the Ether, creating pathways that focus and amplify potentialities.

   - *Example:* In a group meditation session, collective expectation creates a coherent energy field that influences the surrounding environment.

3. **Expectation as a Catalyst:**

   - Expectation accelerates the process of manifestation by enhancing the Ether's responsiveness to thought and intention.

   - *Example:* A researcher expecting a breakthrough aligns their actions and the subtle energy of the Ether to facilitate innovative insights.

Mechanics of Expectation Within the Ether

1. **Wave Interference and Amplification:**

   - Expectation generates vibrational waves that interact with existing patterns in the Ether. Constructive interference amplifies aligned outcomes, while destructive interference diminishes conflicting ones.

- *Example:* A leader's positive expectation for team success reinforces the group's shared energy, amplifying cohesion and productivity.

2. **Expectation and Potential Collapse:**

    - The Ether holds infinite potentialities, and expectation acts as a selective force, collapsing potentials into tangible realities.

    - *Example:* A writer expecting to complete their novel channels energy from the Ether into focused creativity, transforming possibility into accomplishment.

3. **Temporal Influence of Expectation:**

    - Expectation not only affects immediate outcomes but also influences the future trajectory of events.

    - *Example:* A visionary expecting long-term societal change inspires incremental steps that ripple through time, creating a lasting legacy.

Practical Techniques for Engaging the Ether Through Expectation

1. **Visualization with Emotional Resonance:**

    - Combine clear mental imagery with strong emotions to imprint expectation onto the Ether.

    - *Practice:* Visualize your desired outcome with as much sensory detail as possible, infusing the image with genuine belief and joy.

2. **Intentional Affirmations:**

    - Use affirmations to reinforce expectation and align your energy with the Ether's potentialities.

    - *Practice:* Repeat statements like, "The universe supports my goals," while focusing on the feeling of alignment.

3. **Group Expectation and Collective Influence:**

    - Harness the power of shared expectation to amplify impact within the Ether.

    - *Practice:* Organize group intentions, such as collective prayers or meditations, to create a unified field of expectation.

4. **Daily Alignment Practices:**

    - Incorporate routine activities that reinforce coherence between expectation and the Ether.

- *Practice:* Spend 5 minutes each morning setting intentions for the day, visualizing success and alignment.

Examples of Expectation and the Ether in Action

1. **Healing Practices:**

    - Practitioners expecting positive outcomes for their patients channel energy through the Ether, supporting recovery.

    - *Example:* Energy healers using Reiki techniques often report enhanced results when both healer and patient share aligned expectations.

2. **Scientific Phenomena:**

    - Experiments in random number generation have demonstrated that expectation influences outcomes, suggesting interaction with the Ether.

    - *Example:* A group focusing on shifting random outputs consistently skews results, reflecting the power of collective expectation.

3. **Cultural Movements:**

    - Societal shifts often arise from collective expectations that ripple through the Ether to inspire change.

    - *Example:* The expectation of equality during the civil rights movement created a powerful energy field that manifested systemic transformation.

Challenges in Harnessing Expectation and the Ether

1. **Dissonance Between Thought and Expectation:**

    - Misaligned thoughts weaken the coherence of expectation in the Ether.

    - *Solution:* Regularly align thoughts with clear and consistent expectations.

2. **Impatience and Overcorrection:**

    - Unrealistic timeframes can disrupt the natural flow of expectation in the Ether.

    - *Solution:* Trust in the timing of manifestation, reinforcing belief without forcing results.

3. **Competing Energies:**

    - External influences may interfere with individual expectations, diluting their impact.

- *Solution:* Strengthen personal coherence and build resilience to maintain alignment.

Conclusion

Expectation, as an active force within the Ether, demonstrates the profound interconnectedness of thought, intention, and universal energy. By understanding and mastering this interaction, individuals and groups can create meaningful and impactful changes in their lives and the world. The next section will delve into how the interaction between expectation, thought, and intention shapes the tangible realities we experience, completing this exploration of expectation within the UFT framework.

7.4: Connecting Expectation to Reality

Subtitle: How Expectation Shapes Tangible Outcomes

Introduction

Expectation serves as the critical bridge between the abstract and the tangible, transforming thought and intention into reality. Within the Unified Field Theory (UFT), expectation operates as an amplifying and aligning force, shaping outcomes by influencing the probabilities within the Ether. This section explores how expectation translates into real-world manifestations and provides practical strategies to harness its power effectively.

The Process of Translating Expectation into Reality

1. **Initiation Through Thought and Intention:**

   - Thought provides the blueprint, and intention directs energy toward a specific outcome.

   - *Example:* An artist conceptualizes a painting (thought) and dedicates time and effort to creating it (intention).

2. **Expectation as the Catalyst:**

   - Expectation solidifies potentialities, encouraging coherence between intention and manifestation.

   - *Example:* The artist's belief in their ability to complete the painting sustains motivation and focus, turning potential into a tangible creation.

3. **Feedback Loops of Reality Formation:**

   - Every small manifestation reinforces belief, amplifying the energy of expectation and accelerating further results.

- *Example:* A student excelling in one exam gains confidence (feedback), reinforcing their expectation to perform well in future tests.

Mechanisms Linking Expectation to Tangible Outcomes

1. **Vibrational Resonance:**

    - Expectation aligns vibrational frequencies between the individual and the desired outcome within the Ether.

    - *Example:* Entrepreneurs with positive expectations about their ventures often attract investors and collaborators who resonate with their vision.

2. **Coherence and Synchronization:**

    - Aligned thoughts, intentions, and expectations create coherence, synchronizing external events to match internal desires.

    - *Example:* A scientist expecting a breakthrough notices patterns in data that others may overlook, leading to discovery.

3. **Energetic Amplification:**

    - Strong, focused expectation amplifies the energy in the Ether, increasing the probability of manifestation.

    - *Example:* A community collectively expecting positive change influences local policies and initiatives, creating tangible improvements.

Practical Strategies to Bridge Expectation and Reality

1. **Clarity in Vision:**

    - Define your desired outcomes with precision to align expectation with reality.

    - *Practice:* Write a detailed description of your goal, focusing on both measurable results and emotional resonance.

2. **Reinforcement Through Action:**

    - Support expectation with consistent, intentional actions.

    - *Practice:* Break goals into actionable steps and celebrate small victories to sustain positive momentum.

3. **Cultivating Emotional Alignment:**

    - Maintain emotions that match the frequency of your desired outcome.

- *Practice:* Use gratitude exercises to embody the feeling of already achieving your goal.

4. **Harnessing Group Expectation:**
   - Collaborate with others to amplify collective energy and coherence.
   - *Practice:* Join or create mastermind groups where shared expectations reinforce individual efforts.

Real-World Applications of Connecting Expectation to Reality

1. **Personal Development:**
   - Individuals expecting success in personal endeavors consistently achieve higher levels of fulfillment and performance.
   - *Example:* A runner visualizing a race victory improves focus and stamina, enhancing their chances of winning.

2. **Business and Innovation:**
   - Companies led by visionaries with strong expectations for success often outperform their competitors.
   - *Example:* Tech startups expecting global impact align their actions and attract resources to scale rapidly.

3. **Community and Social Movements:**
   - Unified expectation within communities drives collective action and systemic change.
   - *Example:* Grassroots organizations expecting policy reform mobilize efforts that resonate across social and political spheres.

Overcoming Challenges in Manifestation

1. **Misaligned or Negative Expectations:**
   - Counterproductive expectations dilute coherence and hinder outcomes.
   - *Solution:* Regularly assess and realign expectations to match desired goals.

2. **Impatience and Resistance:**
   - Impatience disrupts the natural flow of manifestation, creating frustration and misalignment.
   - *Solution:* Practice mindfulness and trust in the timing of outcomes.

3. **External Influences and Doubts:**

    - External skepticism or negative feedback can weaken personal expectations.

    - *Solution:* Strengthen inner resilience and maintain focus on positive reinforcement.

Examples of Tangible Manifestations

1. **Healing Through Expectation:**

    - Patients expecting recovery experience faster and more comprehensive healing.

    - *Example:* Studies show that optimistic patients recover from surgery more quickly than pessimistic ones.

2. **Technological Innovations:**

    - Innovators expecting breakthroughs align their efforts and resources to achieve transformative outcomes.

    - *Example:* The development of renewable energy technologies has been driven by collective expectation for a sustainable future.

3. **Cultural Shifts:**

    - Societal expectations for equality and justice manifest as tangible policy reforms and cultural evolution.

    - *Example:* Movements for women's rights were fueled by collective expectations for change, leading to global progress.

Conclusion

Expectation is the powerful force that transforms potential into reality, creating a dynamic bridge between thought, intention, and manifestation. By mastering the principles of expectation, individuals and communities can actively shape their lives and the world around them. With this understanding, we complete the exploration of expectation within the Unified Field Theory and transition into the next chapter, where we will explore its integration into a broader societal framework.

Chapter 8: A New Paradigm for Reality

Subtitle: Harnessing Expectation for Individual and Universal Harmony

8.1 Humanity's Role in Universal Evolution

How expectation advances collective consciousness

Introduction

Humanity plays a unique role in the cosmos as both creators and receivers of reality. Through expectation, humans not only shape their individual lives but also influence the collective consciousness that defines the trajectory of universal evolution. This section explores how aligning individual and collective expectations advances humanity's role in the ongoing process of cosmic harmony.

Expectation as a Driver of Collective Evolution

1. **Aligning with Cosmic Laws:**

   - Humanity's expectations influence the collective trajectory, aligning societal progress with universal laws.

   - *Example:* Movements for equality and justice arise from shared expectations for fairness, reflecting humanity's alignment with balance and harmony.

2. **Expectation as Collective Consciousness:**

   - Individual expectations coalesce into a collective consciousness that shapes societal norms and universal resonance.

   - *Example:* Global awareness of climate change has fostered collective expectations for sustainability, driving international action.

3. **Expectation and Conscious Co-Creation:**

   - As humanity becomes more aware of its role in shaping reality, expectations transform into deliberate acts of creation.

   - *Example:* Visionary leaders inspire shared goals, leading to innovations in technology, governance, and art.

Elevating the Role of Humanity Through Expectation

1. **Amplifying Positive Intentions:**

   - When individuals align their expectations with universal harmony, the collective benefits exponentially.

   - *Practice:* Encourage mindfulness practices that focus on positive outcomes, fostering collective coherence.

2. **Overcoming Division Through Shared Expectations:**

- Unifying expectations across cultures and ideologies fosters collaboration and progress.
- *Example:* Global humanitarian initiatives are fueled by shared expectations for peace and prosperity.

3. **Harnessing the Ripple Effect:**

- Individual expectations act as catalysts, creating ripples that influence broader societal changes.
- *Example:* Social media campaigns for change amplify individual voices, creating worldwide movements.

Conclusion
Humanity's role in universal evolution is magnified through the power of expectation. By aligning individual and collective intentions with cosmic principles, humanity actively participates in shaping a future defined by harmony and progress.

8.2 Beyond the Physical: The Metaphysical Implications

How expectation bridges the gap between science and spirituality

Introduction
Expectation transcends the boundaries of physical reality, offering a bridge between science and spirituality. By understanding expectation as a universal force, humanity can harmonize empirical knowledge with metaphysical principles, unlocking a new era of integration and understanding.

Expectation as a Bridge Between Realms

1. **The Scientific Basis of the Metaphysical:**

- Expectation operates within the Ether, linking the physical and metaphysical through quantifiable interactions.
- *Example:* Experiments demonstrating thought's impact on quantum systems reveal the tangible effects of expectation.

2. **Spiritual Principles in Action:**

- Ancient spiritual teachings align with modern understandings of expectation as a creative force.
- *Example:* Practices like prayer and visualization embody expectation's role in shaping reality.

3. **Integration of Science and Spirituality:**

    - The UFT and expectation provide a framework for merging scientific rigor with spiritual wisdom.

    - *Example:* The Golden Ratio (ϕ) serves as a universal constant connecting physical structures with spiritual harmony.

Practical Applications of Metaphysical Expectation

1. **Healing and Energy Work:**

    - Expectation amplifies the effects of healing modalities, creating coherence in the body's energy systems.

    - *Example:* Reiki and meditation practitioners report enhanced outcomes when focused expectations are present.

2. **Manifestation Practices:**

    - Techniques like affirmations and vision boards leverage expectation to materialize desired outcomes.

    - *Practice:* Daily visualization exercises aligned with universal principles enhance clarity and coherence.

3. **Transcending Physical Limitations:**

    - Understanding expectation as a universal force empowers individuals to transcend perceived limitations.

    - *Example:* Athletes and performers achieve peak states by aligning expectations with peak outcomes.

Conclusion
Expectation bridges the gap between the seen and unseen, integrating the physical and metaphysical realms. By embracing this principle, humanity advances not only scientifically but also spiritually, creating a unified understanding of existence.

8.3 The Legacy of Expectation

Shaping a future where humanity harmonizes with universal laws

Introduction
The legacy of expectation is humanity's gift to the cosmos—a testament to the power of alignment and coherence. By embracing expectation as a universal force, humanity can shape a future defined by harmony, sustainability, and purpose. This

section examines how expectation creates a lasting impact on individuals, societies, and the universe itself.

Building a Legacy Through Expectation

1. **Harmonizing with Universal Laws:**

   - By aligning with the UFT, humanity ensures its legacy resonates with the principles of balance and harmony.

   - *Example:* Societal systems designed with ϕ as a guiding principle promote equity and sustainability.

2. **Inspiring Future Generations:**

   - Expectation-driven progress leaves a roadmap for future generations to build upon.

   - *Example:* Educational systems rooted in expectation empower children to shape reality with purpose.

3. **Expanding Humanity's Presence:**

   - Expectation drives exploration and discovery, expanding humanity's influence across the cosmos.

   - *Example:* Space exploration reflects humanity's collective expectation for survival and discovery.

Ensuring a Positive Legacy

1. **Cultivating Collective Responsibility:**

   - Shared expectations for harmony create a foundation for ethical progress.

   - *Practice:* Foster global dialogues that align diverse cultures with universal principles.

2. **Embedding Expectation in Institutions:**

   - Institutions aligned with expectation serve as pillars of progress and stability.

   - *Example:* Governments and corporations adopting harmonic principles promote long-term well-being.

3. **A Legacy of Co-Creation:**

   - Humanity's ultimate legacy is its active participation in the co-creation of reality.

- *Example:* Global efforts to combat climate change demonstrate the power of unified expectation.

Conclusion
The legacy of expectation is humanity's enduring contribution to the cosmos. By harmonizing individual and collective aspirations with universal laws, humanity creates a future defined by balance, progress, and infinite potential.

Conclusion: The Infinite Potential of Expectation

Subtitle: Aligning Humanity with the Universal Symphony

Introduction
As we reach the culmination of this exploration into expectation, one truth emerges: expectation is not merely a human phenomenon but a universal principle. It connects the seen and unseen, the physical and metaphysical, and the individual with the collective. By understanding and aligning expectation with the principles of the Unified Field Theory (UFT), humanity unlocks its infinite potential.

This conclusion reflects on the journey we have taken and charts a path forward—a future where expectation becomes the bridge to universal harmony, personal fulfillment, and global progress.

A Journey Through Expectation

1. **Rediscovering Expectation's Power:**

    - From understanding expectation's mathematical framework to its metaphysical implications, this book has revealed how expectation shapes reality at every level.

2. **Harmonizing with Universal Laws:**

    - Expectation is both the tuning fork and the melody in the cosmic symphony. By aligning with principles like the Golden Ratio (\phi) and coherence, humanity becomes an active participant in the universal order.

3. **Integrating Science, Metaphysics, and Action:**

    - We have seen how expectation bridges disciplines, from governing scientific innovation to inspiring spiritual practices, creating a unified understanding of existence.

The Responsibility of Expectation

1. **Individual Responsibility:**

    • Each person holds the power to align their expectations with universal harmony. By cultivating coherence and clarity, individuals contribute to the collective evolution of humanity.

2. **Collective Responsibility:**

    • Shared expectations drive societal progress, uniting diverse cultures and ideologies under common goals. Humanity's greatest advancements emerge when collective expectations align with universal principles.

3. **Universal Responsibility:**

    • Humanity's role extends beyond Earth. Through expectation, we shape our relationship with the cosmos, ensuring our actions resonate with the balance and sustainability of the universe.

The Vision for the Future

1. **A World Transformed by Expectation:**

    • Imagine a world where individuals master their thoughts, societies function in harmony with universal laws, and technology amplifies humanity's highest aspirations.

2. **Exploring New Frontiers:**

    • From space exploration to advancements in artificial intelligence, the principles of expectation guide humanity toward infinite possibilities.

3. **A Legacy of Unity and Harmony:**

    • The ultimate legacy of expectation is a world where balance, creativity, and connection define human progress.

A Call to Action

This book is not an end but a beginning. The principles of expectation outlined here invite action:

• Reflect on how your personal expectations shape your reality.

• Collaborate with others to align collective expectations with universal harmony.

• Innovate, explore, and create, guided by the principles of coherence and balance.

Let us take this understanding forward—into our lives, our societies, and the cosmos. By aligning expectation with the UFT, we become architects of a future that embodies the infinite potential of humanity and the universe.

Final Words

Expectation is the bridge between who we are and who we can become. It is the force that propels humanity toward the stars, that inspires progress, and that harmonizes our existence with the universal symphony.

The future is not something we merely step into—it is something we shape, consciously and collectively. Let us choose to shape it with intention, clarity, and harmony, ensuring that our expectations resonate eternally within the fabric of reality.

www.ingramcontent.com/pod-product-compliance
Lightning Source LLC
Chambersburg PA
CBHW082253220526
45469CB00009B/2987